Biology

Applying Knowledge and Skills

Writing team
James Torrance
James Fullarton
Clare Marsh
James Simms
Caroline Stevenson

Diagrams by
James Torrance

HODDER
GIBSON
AN HACHETTE UK COMPANY

Cover photo: The crab has been temporarily submerged by a wave from the incoming tide. It is expelling excess sea water from its gills as two jets. Photo courtesy of James Torrance

Photo credits

Fotolia pp. 1, 29 and 63 (background) and running head images. All other photos by James Torrance.

Every effort has been made to trace all copyright holders, but if any have been inadvertently overlooked the Publishers will be pleased to make the necessary arrangements at the first opportunity.

Although every effort has been made to ensure that website addresses are correct at time of going to press, Hodder Gibson cannot be held responsible for the content of any website mentioned in this book. It is sometimes possible to find a relocated web page by typing in the address of the home page for a website in the URL window of your browser.

Hachette UK's policy is to use papers that are natural, renewable and recyclable products and made from wood grown in sustainable forests. The logging and manufacturing processes are expected to conform to the environmental regulations of the country of origin.

Orders: please contact Bookpoint Ltd, 130 Milton Park, Abingdon, Oxon OX14 4SB. Telephone: (44) 01235 827720. Fax: (44) 01235 400454. Lines are open 9.00–5.00, Monday to Saturday, with a 24-hour message answering service. Visit our website at www.hoddereducation.co.uk. Hodder Gibson can be contacted direct on: Tel: 0141 848 1609; Fax: 0141 889 6315; email: hoddergibson@hodder.co.uk

First published in 2013 by
Hodder Gibson, an imprint of Hodder Education,
An Hachette UK Company
2a Christie Street
Paisley PA1 1NB

Impression number	5	4	3	2	1
Year	2017	2016	2015	2014	2013

ISBN: 978 1 4441 9776 1

Illustrations by James Torrance

Typeset in Minion Regular 11/14 by Integra Software Services Pvt. Ltd., Pondicherry, India

Printed in Dubai

A catalogue record for this title is available from the British Library

Contents

Preface

This book has been written specifically to complement the textbook *National 5 Biology*. It provides a wealth of further questions and practice material that correspond with the core text, chapter by chapter. Its contents allow students to practise applying their knowledge and skills, an essential part of the course. Answers to all of the questions are provided.

Unit **1**

Cell Biology

1 Cell structure

1 a) Match cells W, X, Y and Z shown in Figure 1.1 with the following cell types: *Elodea* leaf cell, onion epidermal cell, rhubarb epidermal cell, yeast cell. (3)

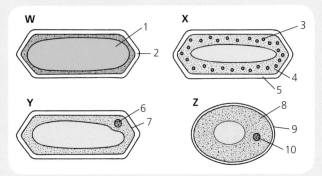

Figure 1.1

   b) Name cell structures 1 to 10. (10)
2 Present the information given in Table 1.1 as a bar chart. (3)

Cell type	Cell length or diameter (micrometres)
Onion epidermis	150
Human egg	100
Sea urchin egg	70
Paramecium	50
Human liver	20
Yeast	8
Human red blood corpuscle	7
Bacillus bacterium	3

Table 1.1

3 Figure 1.2 shows a three-dimensional view of a plant cell with a 'window' cut out to reveal some of its internal structures. Imagine that the cell has been sliced across at planes A, B and C. Match these with views 1, 2 and 3 of the cell's internal structure shown in Figure 1.3. (2)

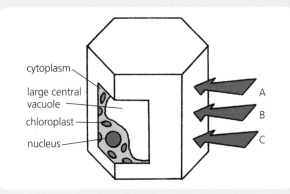

Figure 1.2

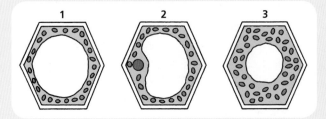

Figure 1.3

4 Two purposes of brown iodine solution in a biological laboratory are:

● to stain structures (if a structure absorbs more iodine solution than its surroundings, it becomes a darker brown colour)
● to test for starch (if starch is present, it turns blue-black in the presence of iodine solution).

Figure 1.4 shows three types of plant cell before and after the addition of iodine solution.

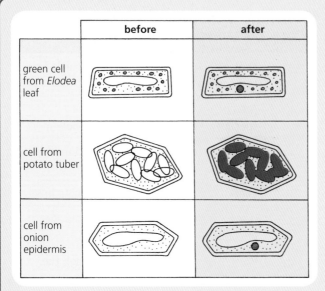

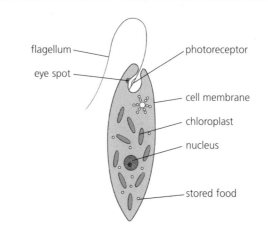

Figure 1.5

Figure 1.4

a) i) Which cellular structure became stained brown?
ii) In which cell types was the process of staining observed in this experiment? (2)
b) i) Which cell type contains starch grains?
ii) Give a reason for your answer. (2)

5 Read the passage and answer the questions that follow.

Euglena viridis, shown in Figure 1.5, is a unicellular organism found living in stagnant pond water. It swims by means of its long whip-like flagellum, movements of which draw the organism forwards. It is able to feed both by photosynthesis and by taking in organic substances present in the water. Its eye spot and photoreceptor are used to guide it towards light.

a) What is the function of *Euglena*'s flagellum? (1)
b) *Euglena* is often found living in water containing rotting organic matter. Suggest why. (1)
c) Why is it of benefit to *Euglena* to be able to swim towards light? (1)
d) Identify ONE structural feature possessed by *Euglena* that is often found in plant cells but never in animal cells. (1)
e) What structural feature typical of all plant cells does *Euglena* lack? (1)
f) In what way does *Euglena* contradict the idea that every living thing can definitely be classified as either a plant or an animal? (2)

6 Figure 1.6 shows a three-dimensional version of a cell with much of the front half cut away. Imagine that the remaining front part is now also cut off. Make a simple, labelled diagram to show the possible appearance of the inside of the cell. (5)

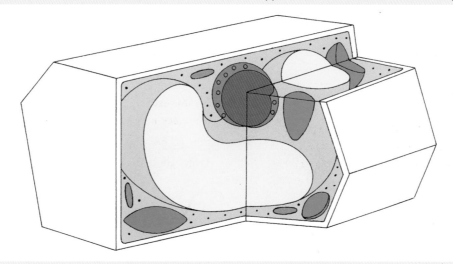

Figure 1.6

7 a) Steps A–F in Table 1.2 give the procedural steps adopted to use the microscope shown in Figure 1.7. Arrange the steps into the correct order starting with D. (1)

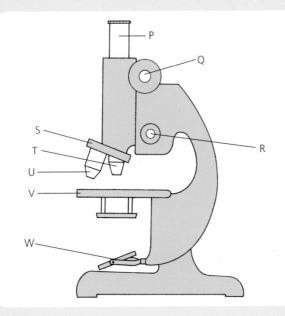

Figure 1.7

Step	Instruction for carrying out step in procedure
A	Look through P and bring the specimen into focus using Q and R.
B	Arrange the prepared slide with the specimen above the centre of the hole in V.
C	Change to high power by turning S until U is above the specimen.
D	Check that T is exactly above the hole in V.
E	Use Q to lower T to about 5 mm from the prepared slide.
F	Move W until light passes up through the microscope.

Table 1.2

b) i) A girl set up this type of microscope to view onion cells but she found that the entire image was very dark. Suggest the step in the procedure that she should check.

ii) A boy found that one half of the image was brightly illuminated and the other half was in darkness. Suggest a remedy for this problem from the list of procedural steps.

iii) When looking at onion cells, a girl found that the image was spoiled by a black mark which, unlike the image of the cells, revolved when she turned P (the eyepiece). Suggest a solution to this problem (not given in Table 1.2). (3)

c) If P contains a lens with a magnification of ×15 and T and U give a magnification of ×8 and ×40, respectively, which of the following is the highest magnification possible for this microscope? (1)

A ×55

B ×120

C ×600

D ×4800

2 Transport across cell membranes

1 Figure 2.1 shows an experiment set up to investigate the process of diffusion. After 10 minutes, a distinct blue-black colour was found to be present inside the Visking tubing sausage but not outside it.

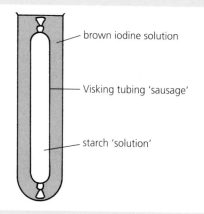

- brown iodine solution

- Visking tubing 'sausage'

- starch 'solution'

Figure 2.1

a) Rewrite the following sentence choosing the correct word from the underlined choice: Iodine solution has reacted with starch <u>inside</u>/<u>outside</u> the Visking tubing 'sausage'. (1)

b) It was concluded from the experiment that molecules of one of the two liquids were small enough to diffuse through the Visking tubing membrane. Identify this liquid. (1)

c) It was concluded that molecules of one of the liquids were too large to diffuse through the membrane. Identify this liquid. (1)

d) An alternative explanation of the results is that the membrane allows molecules of any size to pass in but not out.

   i) Describe the experiment that you would set up to investigate this theory.

   ii) Explain how you would know from your results whether the theory is true or false. (3)

2 A sample of epidermal cells from a red onion was mounted in a sugar solution equal in water concentration to the cell sap and examined to establish the normal appearance of the cells (see Figure 2.2). The cells were then immersed for 10 seconds in solution 1 and re-examined. This procedure was repeated using different concentrations of sugar solution. The appearance of the cells at each stage is shown in the diagram.

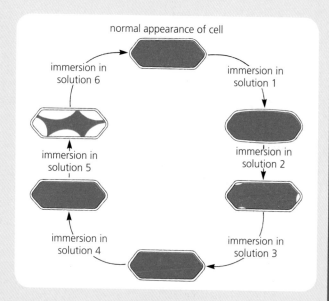

normal appearance of cell

immersion in solution 6

immersion in solution 1

immersion in solution 5

immersion in solution 2

immersion in solution 4

immersion in solution 3

Figure 2.2

a) Which of the five solutions had the

   i) highest water concentration relative to the others?

   ii) lowest water concentration relative to the others? (2)

b) Identify the process occurring at each numbered immersion in Figure 2.2 using the appropriate letter from Table 2.1. (6)

Process	Overall movement of water molecules
X	Movement out of cell exceeds movement in
Y	Movement into cell exceeds movement out
Z	Movement into cell equals movement out

Table 2.1

3 Five thin cylinders were cut out of a potato using a cork borer. The cylinders were trimmed to a length of 50 mm and immersed in sugar solutions of different concentration. The cylinders' lengths were re-measured after 24 hours and the results recorded in Table 2.2.

Concentration of sugar solution (M)	Initial length of cylinder (mm)	Final length of cylinder (mm)	Change in length (mm)	Percentage change in length
0.1	50	58	+8	+16
0.2	50	53		
0.3	50	48		
0.4	50	43		
0.5	50	38		

Table 2.2

a) Copy and complete the last two columns in Table 2.2. (2)
b) Using the axes shown in Figure 2.3, draw a line graph of the percentage change in length of cylinder against molar concentration of sugar solution. (Show gains above and losses below the *x*-axis.) (4)

c) i) Did 0.1 M sugar solution have a higher or lower water concentration compared with potato cell sap?
   ii) Explain your answer. (2)
d) From your graph, estimate the concentration of sugar solution to which potato cell sap is equal in water concentration. (1)
e) i) Would 0.6 M sugar solution have a higher or lower water concentration compared with potato cell sap?
   ii) Explain your answer. (2)
f) Why must the same diameter of cork borer be used to cut out all the cylinders? (1)

4 A type of microscopic organism lives in the estuary of the river shown in Figure 2.4. Two high tides and two low tides occur daily and these affect the organism's volume as indicated by the graph in Figure 2.5.

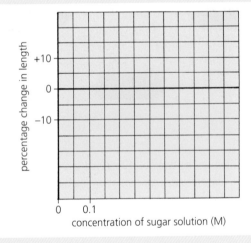

Figure 2.3

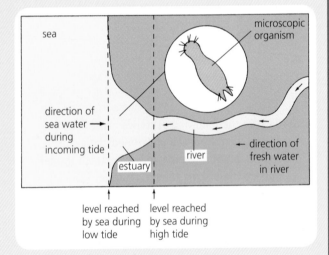

Figure 2.4

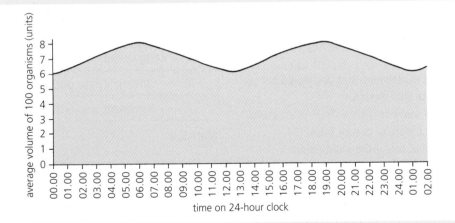

Figure 2.5

a) Rewrite the following sentences, choosing the correct word at each underlined choice.
   i) Compared with sea water, river water has a higher/lower salt concentration.
   ii) When the tide is high, the salt concentration in the river estuary will increase/decrease.
   iii) When the tide is low, the salt concentration in the river estuary will increase/decrease. (3)
b) i) Between which TWO sets of times of day was the animal gaining water by osmosis?
   ii) Was the tide coming in or going out during these times? (2)
c) i) Between which TWO sets of times was the animal losing water by osmosis?
   ii) Was the tide coming in or going out during these times? (2)
d) The data in the graph refer to the average volume of 100 organisms. Why were so many used? (1)

5 If the plant cells shown in Figure 2.6 remained in contact as shown, then water would pass by osmosis from BOTH:
   A   R to Q and Q to P.
   B   Q to S and R to Q.
   C   P to Q and R to S.
   D   Q to P and Q to R.
   (Choose ONE answer only.) (1)

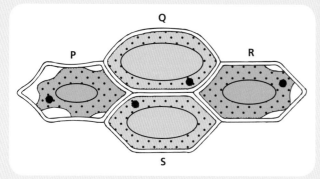

Figure 2.6

6 Figure 2.7 shows two sets of apparatus, A and B, at the start of an experiment. Visking tubing acts as a selectively permeable membrane over a short period of time (for example 2 hours).
   a) Identify the region of higher and lower water concentration in each set-up. (2)
   b) i) Predict the direction in which the level will move in each set-up within the next 30 minutes.
      ii) Explain your answer to i) with reference in each case to the movement of molecules along a concentration gradient. (4)

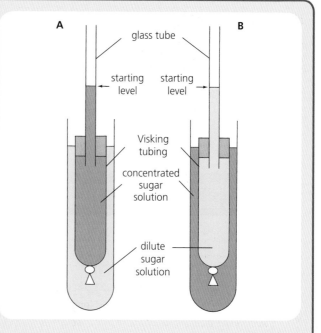

Figure 2.7

   c) What will happen to the rate of movement of the level in tube A if the experiment is repeated using water instead of dilute sugar solution? (1)
7 *Stentor* (see Figure 2.8) is a free-moving, fresh water, unicellular animal that becomes attached by its lower end to a stationary object during feeding.

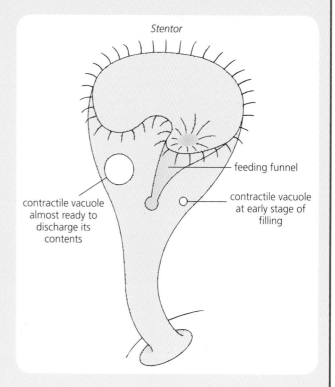

Figure 2.8

In its natural habitat, *Stentor* gains water by osmosis. Unwanted water is removed by two contractile vacuoles. Each filling and emptying of a contractile vacuole is called a pulsation.

In an experiment, specimens of *Stentor* were placed in bathing solutions of different concentrations of salt and viewed under a microscope. Table 2.3 gives mean results from the observation of five animals.

Concentration of salt in bathing solution (%)	Mean number of pulsations/h
0.1	40
0.3	23
0.5	18
0.7	9

Table 2.3

a) What relationship exists between the salt concentration of the bathing solutions and their relative water concentrations? (1)
b) i) What relationship exists between the salt concentration of the bathing solutions and the number of pulsations per hour?
   ii) Explain why. (2)
c) The experiment was repeated using pure water and 1% salt solution. *Stentor*'s contractile vacuoles were found to stop working in one of these liquids.
   i) Identify the liquid in which this happened.
   ii) Predict the response of the animal's contractile vacuoles to the other liquid. (2)
d) i) Identify the variable factor in this experiment involving *Stentor*.
   ii) Name TWO other factors that must be kept constant to make the procedure valid. (3)
e) Why was a mean number of pulsations per hour calculated for each salt concentration? (1)

8 It is possible that some protein carrier molecules actively transport materials against a concentration gradient by rotating within the cell membrane. Rearrange the six stages in Figure 2.9 to give the correct sequence in which this would occur, starting with A. (1)

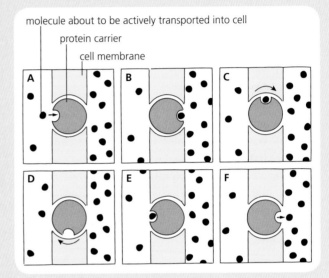

Figure 2.9

9 Table 2.4 refers to the concentrations of certain chemical ions in the cells of *Nitella*, a simple water plant, and in the pond water in which this green plant lives.
   a) Make a generalisation about the concentration of ions in the pond water compared with their concentration in the cell sap. (1)
   b) Calculate the accumulation ratios for potassium and sodium. (1)
   c) In what way do the data support the theory that a cell membrane is **selective** with respect to the process of ion uptake? (1)
   d) i) Do the data support or dispute the suggestion that ion uptake occurs as a result of diffusion?
      ii) Explain your answer. (2)

Substance analysed	Ion concentration of element (mg/l)				
	Calcium (Ca^{2+})	Chloride (Cl^-)	Magnesium (Mg^{2+})	Potassium (K^+)	Sodium (Na^+)
Cell sap	380	3750	260	2400	1988
Pond water	26	35	36	2	28
Accumulation ratio	14.6 : 1	107.1 : 1	7.2 : 1		

Table 2.4

3 Producing new cells

1 Figure 3.1 shows some cells at different stages of mitosis and cell division from a growing root tip.

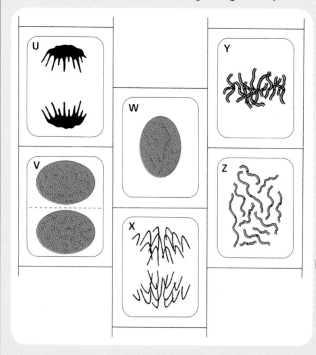

Figure 3.1

a) Arrange them into the correct sequence starting with W. (1)

b) Strictly speaking, which of these are stages in the process of mitosis (nuclear division)? (1)

c) Which row in Table 3.1 correctly matches three of the stages with the events taking place? (1)

	Event		
	Attachment of chromosomes on equator	Separation of chromatids	Division of cytoplasm
A	Y	X	V
B	U	X	V
C	Y	Z	V
D	Y	X	U

Table 3.1

2 Each normal body cell in a kangaroo contains 12 chromosomes.

a) Which lettered part of Figure 3.2 correctly represents two successive cycles of mitosis and cell division? (The numbers refer to the chromosomes present in each cell.) (1)

b) Explain fully your choice of answer. (2)

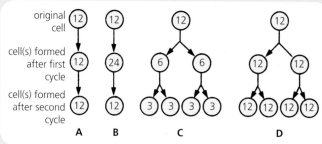

Figure 3.2

3 Read the passage and answer the questions that are based on it.

When a few cells are removed from a living organism and grown in a nutrient solution, the group of cells formed is called a tissue culture. Dividing cells from tissue cultures examined under the microscope show that the time taken by a cell to divide varies from species to species. An onion root tip cell, for example, takes 22 hours to undergo the full cycle whereas a human fibrocyte (a type of connective cell) takes 18 hours from interphase to interphase. Out of this time, only 45 minutes are devoted by the fibrocyte to mitosis. The rest of the time is spent preparing for division of the nucleus.

The rate of cell division is found to be affected by factors such as nutrition and temperature. Cells from a grasshopper embryo require 3.5 hours at 38 °C but 8 hours at 26 °C. Some bacteria divide every 20 minutes at 37 °C but take many hours to do so at low temperatures. Human cells cease to divide at temperatures below 24 °C and above 46 °C.

a) By what means can a supply of cells be readily obtained for studying cell division? (1)

b) A human fibrocyte takes 18 hours to undergo the full cycle of cell division. How much of this time is spent preparing for the actual process of mitosis? (1)

c) Describe the effect of raising the temperature from 26 °C to 38 °C on the rate of division of grasshopper embryo cells. (1)

d) Imagine a single bacterial cell placed in nutrient solution at 37 °C. How many bacteria would be present after 2 hours? (1)

e) Table 3.2 refers to four flasks containing human fibrocyte cells about to be cultured. In which flask(s) would
   i) cell division occur?
   ii) cell division occur at the fastest rate? (2)

Flask	A	B	C	D
Temperature (°C)	17	27	37	47

Table 3.2

f) i) Apart from temperature, what other factor affects the rate of cell division?
   ii) Suggest why. (2)

4 The stages that occur before, during and after mitosis in a certain type of animal cell are represented by the pie chart shown in Figure 3.3. One complete cycle takes 12 hours at 22 °C.
   a) Copy and complete Table 3.3. (2)

Stage	Time taken at 22 °C (hours)
P	
Q	
R	
S	

Table 3.3

b) i) Express as a whole number ratio the total time spent on cell growth to the time spent on production of duplicate chromosomes at 22 °C.
   ii) Express the time spent on mitosis as a percentage of the total time required for one cycle at 22 °C. (2)

c) At 38 °C, the cycle takes 4 hours but the relative time required for each stage remains unchanged. Copy the following sentences choosing only the correct answer from each set of underlined alternatives.

An increase in temperature slows down/fails to affect/speeds up the rate of mitosis and cell division. In this type of cell at 38 °C, stage Q would take 1 hour 20 minutes/1 hour 30 minutes/1 hour 40 minutes and stage S would take 30 minutes/40 minutes/50 minutes. (3)

d) Predict the effect on the rate of mitosis and cell division of lowering the temperature to 10 °C. (1)

5 Figure 3.4 shows a glass spreader. This tool is often used to spread micro-organisms over the surface of a dish of nutrient agar. Before it can be used, it must be sterilised. Figure 3.5 shows six steps (A–F) carried out to sterilise a spreader using burning alcohol. Arrange them into the correct order starting with step B. (1)

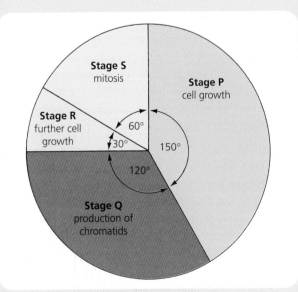

Figure 3.3

Figure 3.4

A

alcohol allowed to burn off

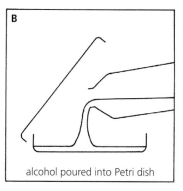

B

alcohol poured into Petri dish

C

some alcohol left on spreader

excess alcohol shaken off spreader and lid placed on Petri dish

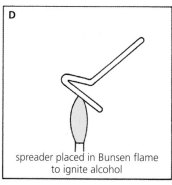

D

spreader placed in Bunsen flame to ignite alcohol

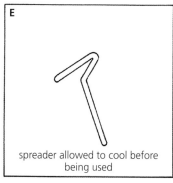

E

spreader allowed to cool before being used

F

spreader dipped in alcohol

Figure 3.5

6 A pupil was asked to find out whether sour milk contains more bacteria than fresh milk. Figure 3.6 shows the procedure that he followed. Spot the THREE mistakes that the boy made and then copy and complete Table 3.4. (6)

Error or omission in procedure	Correct procedure	Reason for following correct procedure

Table 3.4

He collected the following apparatus:

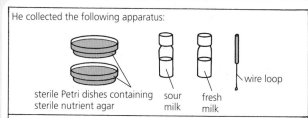

sterile Petri dishes containing sterile nutrient agar | sour milk | fresh milk | wire loop

He held the tip of the wire loop in a Bunsen flame until it was red hot and then cooled it by waving it in the air.

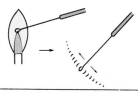

He collected a small drop of sour milk on the loop.

He spread the sour milk over the nutrient agar in one of the plates.

He then used the wire loop to take a sample of fresh milk which he spread in the same way onto the second plate of agar.

He sealed the dishes with sticky tape and put them in a warm incubator.

Figure 3.6

7 Figure 3.7 shows apparatus A set up to find out whether live yeast cells can grow and convert sugar to alcohol. Apparatus B was set up as the control. State SIX ways in which B needs to be changed to make it a valid control. (6)

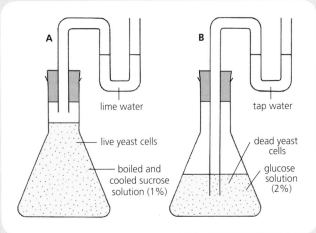

Figure 3.7

8 Yeast cells that are growing and respiring give out bubbles of carbon dioxide. When live yeast is used in bread-making, these carbon dioxide bubbles make the dough rise. Two groups of pupils set up an experiment to investigate the effect of temperature on the action of baker's yeast. Group A made their dough from flour, yeast and water; group B used flour, yeast, sugar and water. Each poured dough into three measuring cylinders up to the $10\,cm^3$ mark and left the cylinders at 5 °C, 20 °C and 35 °C. The graphs in Figure 3.8 show their results.

a) i) Which variable factor did both groups set out to investigate?
   ii) How many different values of this variable factor were used?
   iii) In both cases, what overall effect did the variable factor have on the volume of dough produced? (3)
b) i) By what ONE factor did group A's investigation differ from that of B at the start?
   ii) What overall effect did this factor have on the volume of dough produced?
   iii) Explain your answer to ii). (3)
c) By how many cm^3 was the volume of group B's dough greater than that of group A at 30 minutes and 20 °C? (1)
d) By how many times was the volume of group B's dough greater than that of group A at 50 minutes and 35 °C? (1)
e) i) Predict the results of repeating the experiments at 250 °C (the temperature at which bread is baked).
   ii) Explain your answer to i). (2)

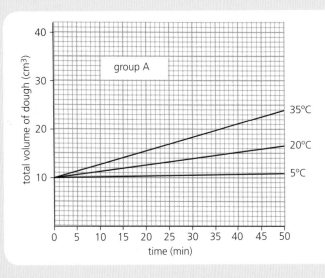

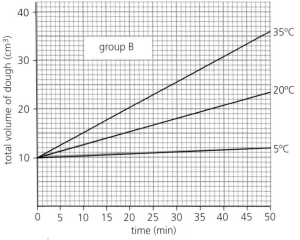

Figure 3.8

4 DNA and the production of proteins

1 a) Match the contents of boxes 1, 2 and 3 in Figure 4.1 with the following descriptions.
  i) a DNA base
  ii) a chromosome
  iii) a small portion of DNA's two strands held together. (2)

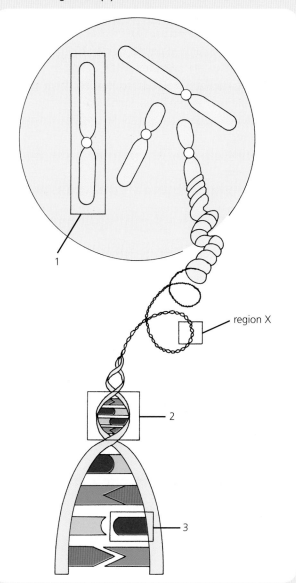

Figure 4.1

b) Suggest why region X in the diagram fails to represent one gene adequately. (1)

2 Figure 4.2 shows a small part of a DNA strand. Draw a simple diagram to show the appearance of the strand of mRNA that is complementary to this DNA strand. (1)

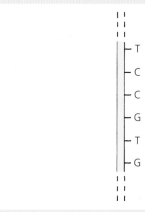

Figure 4.2

3 a) State the number of base G molecules that would be present in a DNA molecule that contains a total of 10 000 base molecules of which 22% are base T. (1)
  b) Calculate the percentage of base A molecules present in a DNA molecule containing 1000 bases of which 300 are base C. (1)

4 Figure 4.3 shows a small part of a DNA molecule where the four types of base molecule are represented by the letters A, T, G and C. Which part of Figure 4.4 supplies the information missing from box X in Figure 4.3? (1)

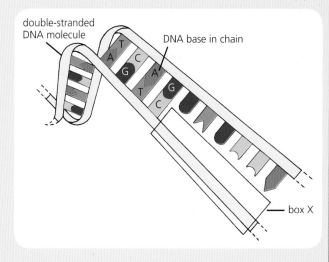

double-stranded DNA molecule

DNA base in chain

box X

Figure 4.3

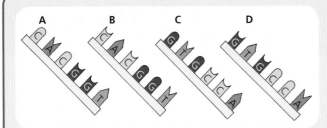

Figure 4.4

5 What is the difference between the DNA of one member of a species and that of another member of the same species? (Choose the ONE correct answer.) (1)
   A The order in which the bases occur in their chromosomes.
   B The type of bases present in their chromosomes.
   C The number of strands that occur in their chromosomes.
   D The type of bonds present between the bases in their chromosomes.

6 The set of results in Table 4.1 shows an analysis of the DNA bases contained in the cells of a cow.

Base composition			
X	G (guanine)	Y	Z
28.2%	21.8%	21.8%	28.2%

Table 4.1

Which row in Table 4.2 is a possible correct identification of the bases? (1)

	X	Y	Z
A	C (cytosine)	A (adenine)	T (thymine)
B	T (thymine)	A (adenine)	C (cytosine)
C	A (adenine)	C (cytosine)	T (thymine)
D	C (cytosine)	T (thymine)	A (adenine)

Table 4.2

7 The building-up of a molecule of protein whose structure is dictated by information held in a chromosome involves the following stages.
   A mRNA passes out of the nucleus and into the cytoplasm of the cell.
   B A region of DNA molecule uncoils and opens up.
   C Amino acids are assembled into protein in a sequence determined by the order of bases on the mRNA.
   D A strand of mRNA is formed as the complement of one of the DNA strands.
   a) Arrange the stages into the correct order. (1)
   b) Where in the cell does stage D occur? (1)
   c) Which cellular structure must be present for stage C to occur? (1)

8 If each amino acid molecule weighs 100 mass units, which ONE of the following is the weight in mass units of a protein molecule synthesised from an mRNA molecule that is 600 bases long? (1)
   A 2000
   B 6000
   C 20 000
   D 60 000

5 Proteins and enzymes

1 Hydrogen peroxide is a chemical that breaks down very slowly into oxygen and water. The experiment shown in Figure 5.1 was set up to investigate the effect of catalase (an enzyme present in living cells) on the rate of this chemical reaction.

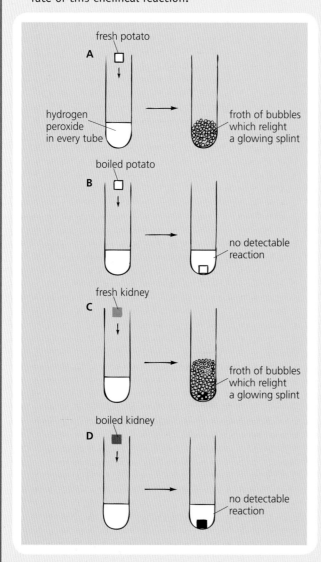

Figure 5.1

a) Identify the substrate in this experiment. (1)
b) i) Which TWO tubes received cells that provided a supply of active catalase?
   ii) Describe the effect that this enzyme has on the rate of breakdown of hydrogen peroxide.

iii) Name ONE of the end products formed and describe how it was identified. (4)
c) Which TWO tubes received denatured enzyme? (1)
d) Explain at molecular level why a denatured enzyme is ineffective. (2)

2 The experiment shown in Figure 5.2 was carried out by a group of pupils to investigate the effect of acid on the action of the enzyme pepsin. Other groups were asked to design a control experiment. Some of their suggestions are shown in Figure 5.3.

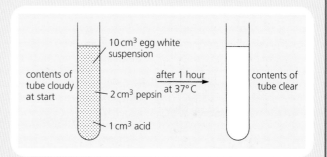

Figure 5.2

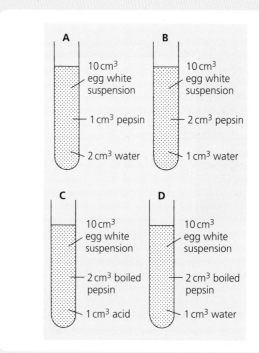

Figure 5.3

a) Which ONE of the four designs is a valid control? (1)

b) Explain, in turn, why each of the others is invalid. (3)

3 In an investigation into the action of the enzyme urease, three versions of the apparatus shown in Figure 5.4 were set up. Urease activity is indicated by the release of ammonia gas which turns red litmus paper blue. The results of the experiment are summarised in Table 5.1.

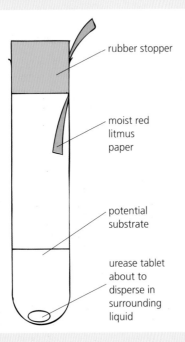

rubber stopper

moist red litmus paper

potential substrate

urease tablet about to disperse in surrounding liquid

Figure 5.4

Test tube	A	B	C
Potential substrate	Cloudy egg white suspension	Urea solution	Starch suspension
Colour of litmus at start	Red	Red	Red
Colour of litmus after 30 minutes	Red	Blue	Red

Table 5.1

a) What was the ONE variable factor investigated in this experiment? (1)

b) Name THREE factors that must be kept constant when setting up tubes A, B and C. (3)

c) Suggest why the three test tubes should be shaken gently at the start of the experiment. (1)

d) i) Which test tube showed evidence of urease activity?

   ii) How could you tell? (2)

e) What characteristic of enzymes does this experiment demonstrate to be true of urease? (1)

4 One gram of chopped liver was added to hydrogen peroxide solution at different pH values and the time taken to collect $1\,cm^3$ of oxygen was noted in each case. The results are given in Table 5.2.

pH of hydrogen peroxide solution	Time to collect $1\,cm^3$ of oxygen (seconds)
6	105
7	78
8	57
9	45
10	52
11	66
12	99

Table 5.2

a) Present the results in the form of a line graph with pH on the horizontal axis. (3)

b) From your graph, state the pH at which the enzyme was
   i) most active
   ii) least active. (2)

c) Of the pH values used in this experiment, which is the optimum for the enzyme present in the liver cells? (1)

d) How could you obtain an even more accurate measurement of the optimum pH at which this enzyme works? (1)

5 The bar graph in Figure 5.5 shows the results from an investigation into the effect of alcohol on the activity of an enzyme that digests food in the small intestine of humans.

a) In general what effect does increasing concentration of alcohol have on the mass of food digested per hour? (1)

b) Which concentration of alcohol has the greatest effect on the activity of the enzyme? (1)

c) With reference to this experiment only, suggest ONE reason why people are advised against drinking excessive quantities of alcohol. (1)

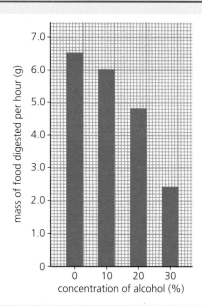

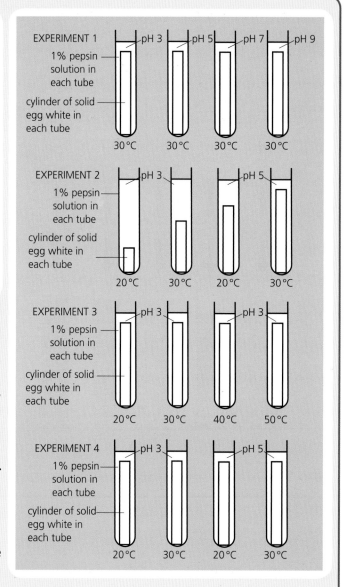

Figure 5.5

d) Express the masses of food digested under conditions of 10, 20 and 30% alcohol as a simple whole number ratio. (1)

e) Calculate the percentage decrease in enzyme activity that occurred between 10 and 20% alcohol. (1)

6 Gastric juice is produced by the lining of the human stomach. It contains hydrochloric acid and pepsin, a protein-digesting enzyme. Egg white is rich in protein. Four experiments were set up as shown in Figure 5.6 in an attempt to investigate the influence of certain factors on the activity of pepsin.

a) In what way would a cylinder of solid egg white become altered if pepsin successfully promoted the digestion of protein? (1)

b) Which of the four experiments is a valid investigation into the influence of temperature on pepsin activity? (1)

c) Which of the four experiments is a valid investigation into the effect of pH on pepsin activity? (1)

d) i) Which experiment really consists of two variable factors being investigated at the same time?

ii) Predict which combination of these two factors would be most effective at promoting the activity of pepsin.

iii) Explain your answer to ii). (4)

e) i) Which experiment attempts to investigate the effect of mass of substrate on enzyme activity?

ii) Why is the set-up invalid as it stands?

iii) Draw a diagram showing the four tubes correctly set up ready to investigate the effect of mass of substrate on pepsin activity. (4)

Figure 5.6

7 Explain each of the following in terms of enzymes:

a) Fevers that raise the body temperature to over 42 °C are normally fatal to human beings. (2)

b) Vinegar (an acid) is used to preserve food against attack by micro-organisms. (2)

c) Cheese kept in a warm room turns mouldy much more quickly than cheese kept in a refrigerator. (2)

8 When iodine solution is added to starch, the starch turns purple (blue-black). Amylase is an enzyme that digests starch in the human mouth and small intestine. Figure 5.7 shows an experiment set up to investigate the effect of aspirin on amylase activity. The aspirin 'solution' was prepared by crushing several aspirin tablets and adding them to 20 cm³ of water.

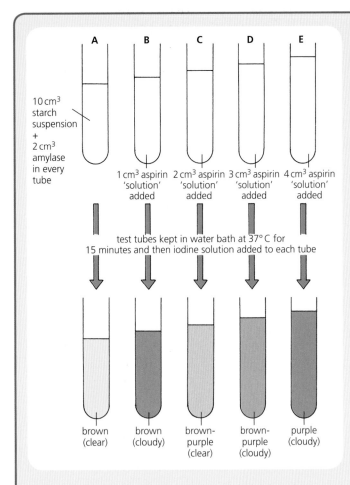

Figure 5.7

Colour	Level of enzyme activity indicated (units)
Purple (cloudy)	0
Brown-purple (cloudy)	1
Brown-purple (clear)	2
Brown (cloudy)	3
Brown (clear)	4

Table 5.3

a) Draw a bar chart of the results using the information in Table 5.3 to convert qualitative values into quantitative ones. (3)

b) i) Explain why the highest level of enzyme activity is indicated by a clear brown colour.
   ii) Explain why the lowest level of enzyme activity is indicated by a cloudy purple colour. (2)

c) What conclusion can be drawn from the results of this experiment? (1)

d) i) Strictly speaking, why can a valid comparison of the results obtained in tubes A–E not be made?
   ii) State how the experiment's design could be improved to eliminate this problem. (2)

6 Genetic engineering

1 The six lettered boxes in Figure 6.1 are stages in the production of human growth hormone (HGH) by an industrial process following the formation of genetically modified (GM) bacteria.

A HGH extracted, concentrated and purified

B HGH gene inserted into suitable bacterium

C bacteria allowed to reproduce rapidly and produce HGH

D gene responsible for HGH identified

E GM bacterium given optimum growing conditions

F section of DNA that contains HGH gene extracted from source chromosome

Figure 6.1

a) Cut up a copy of Figure 6.1 and arrange the boxes into the correct sequence. (1)
b) Briefly describe the means by which stage B could be achieved including reference to an appropriate vector in your answer. (2)
c) Draw a simple labelled diagram to show stage B on completion. (2)

2 Read the passage and answer the questions that follow it.

Lactose is a sugar present in whey (the unwanted remains of milk after the curds have been used for cheese-making). It is composed of two simpler sugars. Brewer's yeast is unable to convert lactose to alcohol because the fungus cannot take the sugar in through its cell membrane. Even if lactose could enter a yeast cell, the organism lacks the enzyme needed to break the bond holding together the components of a molecule of lactose – one molecule of glucose and one of galactose.

However, genetic engineers have solved these problems by transferring genetic material from a rare species of yeast into brewer's yeast. The genetically modified strain formed has the genes for making both lactose permease (the enzyme needed for lactose to enter the cell) and lactase (the enzyme that digests lactose to glucose and galactose). As a result, normal fermentation can proceed using this GM strain.

a) Describe the structure of a molecule of lactose. (1)
b) Give TWO reasons why traditional brewer's yeast is unable to use lactose sugar as a foodstuff. (2)
c) With reference only to the information in the passage, describe what is meant by the term *genetic engineering*. (1)
d) Give TWO reasons why the GM strain of yeast is able to make use of lactose. (2)
e) Which of the following chemical reactions is the GM strain able to bring about? (1)

A   lactase ⟶ alcohol
B   alcohol ⟶ galactose
C   glucose ⟶ lactose
D   glucose ⟶ alcohol

3 Figure 6.2 refers to the planting of three types of genetically modified crop in a part of North America.

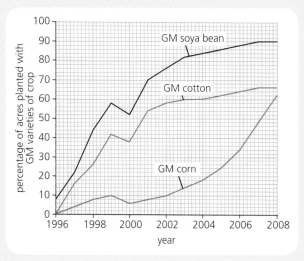

Figure 6.2

a) i) What percentage of acres planted with soya bean in 2003 were of the GM variety?

ii) In which year were 62% of acres planted with cotton occupied by the GM type? (2)

b) Make a generalisation about the trend indicated by the data in the graph. (1)

c) By how many times was the percentage of acres planted with GM corn greater in 2007 compared with 2000? (1)

d) Of which type of crop were GM seeds probably being planted before the period of time covered by the graph? (1)

e) i) In which year did demand for GM seeds outstrip supply?

ii) Explain your answer to i). (2)

4 Read the passage and answer the questions that follow it.

Avian influenza ('bird flu') is caused by a virus able to 'jump species' from wild birds to domestic chickens. In recent times the disease has completely wiped out the livelihood of many poultry farmers in parts of Asia. In addition to being an economic problem, there remains the possibility that bird flu will be transmitted to humans in the future and cause a disease epidemic.

Scientists in the UK have now developed a strain of genetically modified (GM) chickens that become ill if infected with the virus but are unable to transfer the virus to other chickens with which they come in contact. To produce these GM chickens, the scientists inserted a gene into the bird's genetic material (genome). This gene produces a 'decoy' molecule that interferes with the virus's replication process.

No differences have been found to exist between the transgenic chickens and their non-GM relatives. Therefore scientists are confident that if the GM chickens were to be produced on a commercial scale, they would be safe for human consumption. In the future, scientists hope to develop birds that are fully resistant to avian flu.

a) By what means was the strain of transgenic chickens referred to in the passage created? (1)

b) i) Can this new GM strain of chicken catch avian influenza?

ii) Why are the members of this new transgenic strain considered to be better than their non-GM relatives? (2)

c) i) Would the GM chickens be safe to eat?

ii) Justify your answer. (2)

d) Give TWO benefits that would result from the development of transgenic chickens fully resistant to bird flu in the future. (2)

5 Copy and complete Table 6.1 using the following answers. (5)

A Crop is of improved nutritional value

B Crop survives but weeds die when weedkiller is applied

C Production of a chemical that acts as a natural antifreeze

D Production of a protein that acts as a natural insecticide

E Rice

F Shelf life of fruit is extended

Genetically modified (GM) crop plant	Role of inserted gene	Beneficial effect
	Production of chemical that can be converted in human body to vitamin A	
Apple	Blockage of production of chemicals that promote ripening	
Pea		Leaves able to resist attack by caterpillars
Soya	Production of a protein that gives resistance to weedkiller	
Strawberry		Fruit protected against damage by frost

Table 6.1

6 Figure 6.3 shows six members of a pressure group opposed to the practice of genetic engineering and six experts presenting a defence of it.

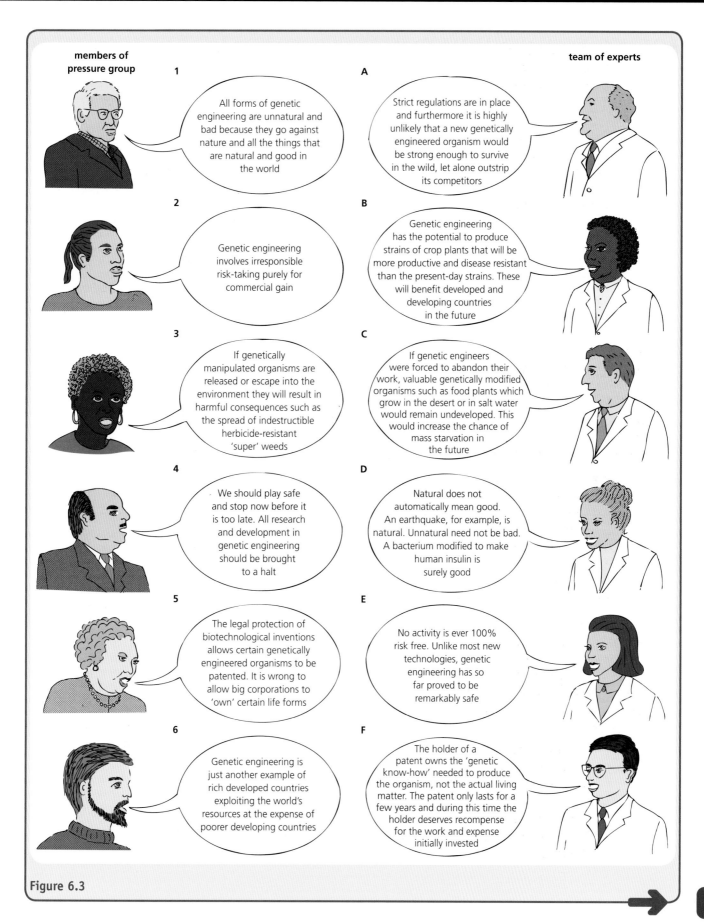

Figure 6.3

a) Match statements 1–6 with replies A–F. (5)

b) Are you in favour of genetic engineering? Justify your answer.

7 Read the passage and answer the questions that follow it.

Fish farming is a rapidly growing segment of agriculture and the global consumption of farmed fish is expected to outpace that of beef in the near future. A genetically modified (GM) type of Atlantic salmon has been developed for use in fish farms in the USA. This type of salmon reaches market size in half the time taken by the wild type because it contains a growth-regulating gene from the Pacific Chinook salmon. In addition, the GM salmon is able to grow all through the year because it contains a promoter gene from the ocean pout – a large eel-like fish. The wild type of Atlantic salmon grows during spring and summer only.

In 2010, after reviewing the evidence, the US Food and Drug Administration advised that the GM Atlantic salmon is as safe a food as conventional wild-type Atlantic salmon and that the former was highly unlikely to affect significantly the environment in any adverse way. However, this advice has been rejected by those who claim that if the GM variety escapes into the wild, it would have the potential to feed more efficiently than the wild type of salmon. Since GM salmon also have the potential to survive for twice as long as the wild type, the protesters argue that the GM salmon could outcompete the wild type in the ocean.

Supporters of the use of GM salmon point out that these fish are poorer swimmers because they have smaller muscle fibres. In addition the males have a low level of reproductive success owing to a low sperm count. They also point out that the fish to be cultivated in fish farms would be sterile females.

a) Briefly describe the means by which the GM Atlantic salmon was developed. (2)

b) i) Identify TWO characteristics of GM salmon that suggest that escapees from fish farms would be able to outcompete wild salmon in the ocean.

   ii) What TWO characteristics of GM salmon support the idea that they would not be able to compete successfully if they escaped from captivity? (4)

c) Explain the advantage of cultivating *sterile* fish in a fish farm. (1)

d) Some people insist that when GM Atlantic salmon goes on sale as a foodstuff, it should carry a label identifying its transgenic origins. Suggest why the manufacturers are opposed to this idea. (1)

7 Respiration

1 Table 7.1 shows the energy content of ten foodstuffs and the percentage of each class of food in the foodstuff.

Foodstuff	Energy content (kJ/g)	Class of food		
		Protein (%)	Fat (%)	Carbohydrate (%)
Chocolate	24.0	9.0	37.0	54.0
Corn oil	38.0	0.0	100.0	0.0
Egg white	19.0	100.0	0.0	0.0
Ice cream	8.0	4.0	11.0	20.0
Lard	37.0	0.0	100.0	0.0
Olive oil	39.0	0.0	100.0	0.0
Peas (boiled)	2.1	5.0	0.0	7.8
Potatoes (boiled)	3.3	1.4	0.0	19.5
Potatoes (chipped)	9.9	4.0	9.0	37.0
Sucrose	19.0	0.0	0.0	100.0

Table 7.1

a) i) Identify the THREE foods in Table 7.1 that are composed exclusively of fat.
   ii) Calculate the mean energy content of fat based on these three foods. (3)
b) Identify the food composed exclusively of:
   i) protein
   ii) carbohydrate.
   iii) For each of these classes of food, state its energy content. (4)
c) Express the energy content of protein:fat:carbohydrate as a simple whole number ratio. (1)
d) By how many times is the energy content of chipped potatoes greater than that of boiled potatoes? (1)

2 Figure 7.1 shows the process of energy transfer in skeletal muscle tissue. Copy the diagram and complete it by adding four arrow heads and supplying the words missing at positions 1 to 6. (8)

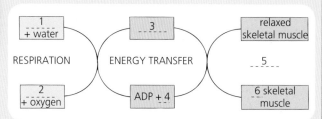

Figure 7.1

3 The graph in Figure 7.2 shows the rate of uptake of oxygen by a plant in darkness at different temperatures.

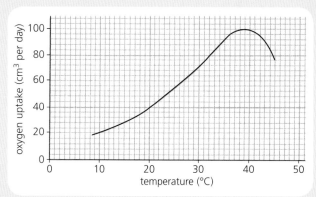

Figure 7.2

a) At what temperature was a volume of 60 cm³ of oxygen per day being taken up? (1)
b) What volume of oxygen was taken up at 15 °C? (1)
c) At what temperature was the highest volume of oxygen taken up? (1)
d) By how many times was the volume of oxygen taken up at 30 °C greater than that taken up at 10 °C? (1)
e) What increase in temperature was required to double the volume of oxygen taken up at 20 °C? (1)

4   Figure 7.3 gives a summary of the chemistry of respiration in a yeast cell.

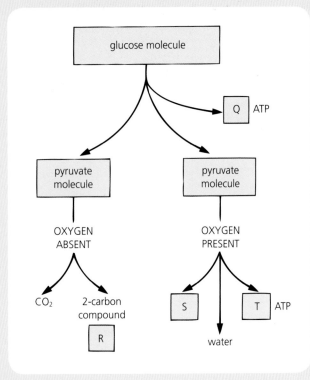

Figure 7.3

a) Identify:
   i) the number of molecules that should have been inserted in boxes Q and T
   ii) the names of the products that should have been inserted in boxes R and S. (4)
b) Predict the fate of substance S. (1)
c) State the number of molecules of ATP gained from:
   i) the partial breakdown of one glucose molecule during fermentation
   ii) the complete breakdown of one glucose molecule during aerobic respiration. (2)
5   In Figure 7.4, graph 1 presents the results of an experiment set up to investigate the effect of oxygen concentration on active uptake of potassium ions and consumption of sugar by root cells of barley seedlings. Graph 2 gives the results from an experiment set up to investigate the effect of temperature on potassium ion uptake by barley roots.
a) **i)** From graph 1, state the effect that an increase in oxygen concentration from 0 to 30% has on rate of ion uptake.

ii) Suggest why ion uptake levels off beyond 30% oxygen.
iii) What relationship exists between units of ion absorbed and units of sugar present in cell sap? Suggest why. (4)

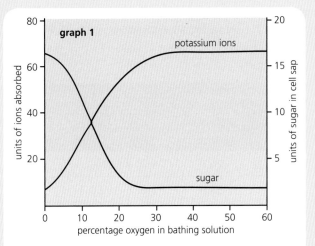

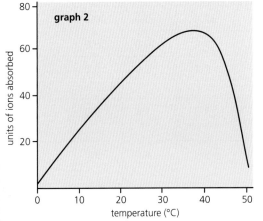

Figure 7.4

b) **i)** State the temperature at which the greatest uptake of ions occurs in graph 2.
   ii) Account for the sudden decline in ion uptake shown by the graph. (2)
6   The following list gives four stages that occur during aerobic respiration. Put them into the correct order.
A   Molecule of pyruvate broken down by enzyme action.
B   Molecule of glucose split into two by enzyme action.
C CO_2 and a large quantity of energy released.
D   Small quantity of energy released. (1)

7 The data in Table 7.2 refer to average values for
   athletes in training for different events.

Athletic event	Volume of oxygen needed for event (l)	Volume of oxygen consumed during event (l)	Oxygen debt (l)	Percentage of energy obtained for event from aerobic respiration	Percentage of energy obtained for event from anaerobic respiration
100 metres	10	0.5	9.5	5	95
800 metres	25	8	17	32	
1500 metres	36	18	18		
10 000 metres	150	135			
marathon (42 195 metres)	700			98	2

Table 7.2

a) Copy and complete the table. (8)
b) Present the data from the two right-hand columns
   of your table as a bar chart. (4)

8 Photosynthesis

1 The apparatus shown in Figure 8.1 was set up to demonstrate that carbon dioxide is necessary for photosynthesis.

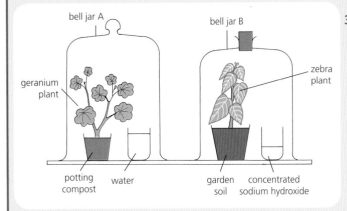

Figure 8.1

a) Name SIX ways in which bell jar B's set-up needs to be altered to make the experiment valid. (6)

b) Once these changes have been made, what TWO environmental factors must be kept constant during the experiment? (2)

2 Table 8.1 refers to the uptake of carbon dioxide by two green plants.

Plant	Total volume of CO_2 entering plant daily (mm³)	Number of leaves	Mean area of one leaf (mm²)
X	24 000	12	500
Y	40 000	5	1000

Table 8.1

a) Calculate the daily rate of diffusion of CO_2 into plant X in mm³ CO_2 per mm² of leaf. (1)

b) By how many times is the daily rate of diffusion of CO_2 into plant Y greater than that into plant X? (1)

3 The graph in Figure 8.2 shows the variation in carbon dioxide concentration of the air surrounding the leaves of a potato crop during 3 days in summer.

a) On which day and between which times (to the nearest hour) did the concentration of CO_2 drop at the fastest rate? (1)

b) Which physiological process was responsible for each decrease in CO_2 concentration in the graph? (1)

c) At approximately what time did:
   i) day break?
   ii) night fall? (2)

d) i) Suggest which day was probably the least sunny.
   ii) Explain your answer. (2)

4 The plant in Figure 8.3 was destarched before being set up as shown. This demonstration really consists of three separate experiments being done on the plant at the same time.

a) After the plant has been in bright light for 3 days, which two discs should be tested for starch in order to show that:
   i) light is essential for photosynthesis?
   ii) chlorophyll is essential for photosynthesis?
   iii) carbon dioxide is essential for photosynthesis?
   iv) Explain your answer in each case. (6)

b) Why is the plant destarched before being used in this experiment? (1)

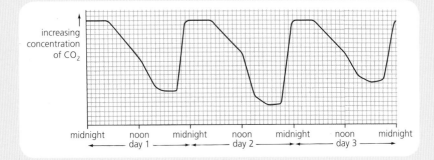

Figure 8.2

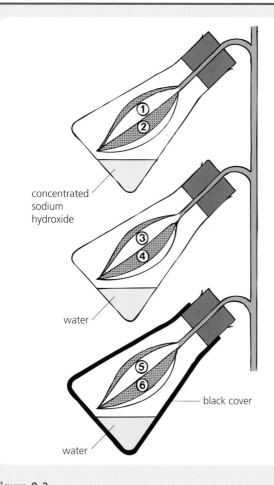

concentrated
sodium
hydroxide

water

black cover

water

Figure 8.3

5 Assume that to produce one unit of sugar by
photosynthesis, each of the plants referred to in
Table 8.2 must receive three units of CO_2, three units
of water and six units of light energy.

Plant	Units of CO_2 available to plant	Units of water available to plant	Units of light available to plant
A	12	12	12
B	6	12	24
C	12	24	12
D	12	12	24
E	24	12	12

Table 8.2

a) How many units of sugar will plant A be able to
make? (1)

b) Which factor is in short supply and is holding up
plant A's photosynthesis? (1)

c) Which plant will be able to make the greatest
number of sugar units? (1)

d) If plant E is given an unlimited supply of light
energy, how many units of sugar will it be able to
make under the conditions given? (1)

6 Figure 8.4 shows the apparatus set up to
investigate the effect of temperature on the rate of
photosynthesis by the water weed
Elodea. The initial level on the
measuring cylinder was recorded
at 14.00 hours on Monday 7
April; the final level was recorded
at 18.00 hours on Friday 11 April.
The apparatus was illuminated
continuously using a powerful lamp.

a) Name the gas released by *Elodea*
that collects in the measuring
cylinder. (1)

b) Calculate the hourly rate of
photosynthesis. (1)

c) i) Describe how the apparatus
should be adapted to measure
rate of photosynthesis at
25 °C, 40 °C and 55 °C.

ii) Explain why an interval of
time must be allowed between
the readings taken at each
temperature. (2)

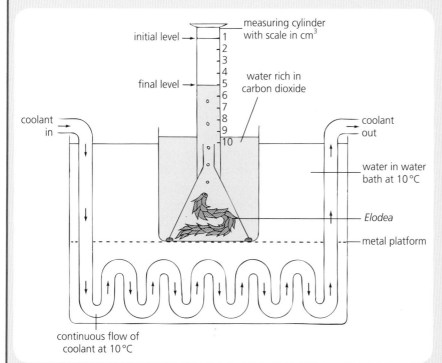

initial level

measuring cylinder
with scale in cm^3

final level

water rich in
carbon dioxide

coolant
in

coolant
out

water in water
bath at 10 °C

Elodea

metal platform

continuous flow of
coolant at 10 °C

Figure 8.4

d) i) Predict at which of the temperatures given in part c) i) the photosynthetic rate would be poorest.

ii) Explain your choice. (2)

e) What is the reason for using water rich in carbon dioxide? (1)

f) Describe how the apparatus could be employed to measure the effect of light intensity on rate of photosynthesis. (2)

7 A horticulturist grew specimens of a species of tomato plant in seven glasshouses each containing a different concentration of carbon dioxide in the air surrounding the plants. After a standard length of time, he measured the mass of sugar produced at each concentration of carbon dioxide, as shown in Table 8.3.

Carbon dioxide concentration (units)	Sugar production (g/kg of dry plant)
0	0.0
10	1.7
20	3.3
30	4.8
40	5.7
50	6.0
60	6.0
70	6.0

Table 8.3

a) i) Which factor is used to measure photosynthesis in this experiment?

ii) What additional information would you need in order to express this factor as the rate of photosynthesis?

iii) Name another factor that could be used to measure the rate of photosynthesis. (3)

b) Express the mass of sugar produced per kilogram of dry plant at a carbon dioxide concentration of 70 units as a percentage of the dry mass of plant material. (1)

c) Using graph paper, construct a line graph of the results. (3)

d) From your graph, state what the sugar production would be at 15 units of carbon dioxide. (1)

e) i) Explain how the graph shows that carbon dioxide was the factor limiting photosynthesis at concentrations of 0 to 50 units.

ii) Explain how the graph shows that carbon dioxide was NOT the limiting factor from a carbon dioxide concentration of 50 units onwards.

iii) Name the factor that could have been limiting the process of photosynthesis from 50 units of carbon dioxide onwards. (5)

8 Table 8.4 shows the results of an investigation where three experiments were carried out simultaneously. A large number of young tomato plants were planted in glasshouses. The glasshouses were subjected to different concentrations of carbon dioxide (CO_2) and different light intensities as indicated in the table. All the other conditions in the glasshouses were kept the same.

Glasshouse	Growth conditions	Mean yield of fruit (kg/plant)	Increase in mean yield of fruit (kg/plant)
1	Normal CO_2 concentration; normal daylight	3.4	–
2	Increased CO_2 concentration; normal daylight	5.7	2.3
3	Normal CO_2 concentration; increased light intensity	8.3	[box X]
4	Increased CO_2 concentration; increased light intensity	[box Y]	6.7

Table 8.4

a) i) Explain why all the conditions in the glasshouses apart from light intensity and carbon dioxide concentration must be kept the same.

ii) Identify TWO of these other conditions. (3)

b) What was the mean yield of fruit when the plants were grown in normal conditions? (1)

c) Supply the information missing from box X. (1)

d) i) What was the increase in mean yield when the plants were given extra CO_2 and extra light?

ii) Supply the information missing from box Y. (2)

e) On its own, which factor had the greater effect on yield of tomatoes – increased CO_2 concentration or increased light intensity? (1)

f) Why was a large number of young tomato plants used? (1)

g) What is the purpose of the procedure carried out in glasshouse 1? (1)

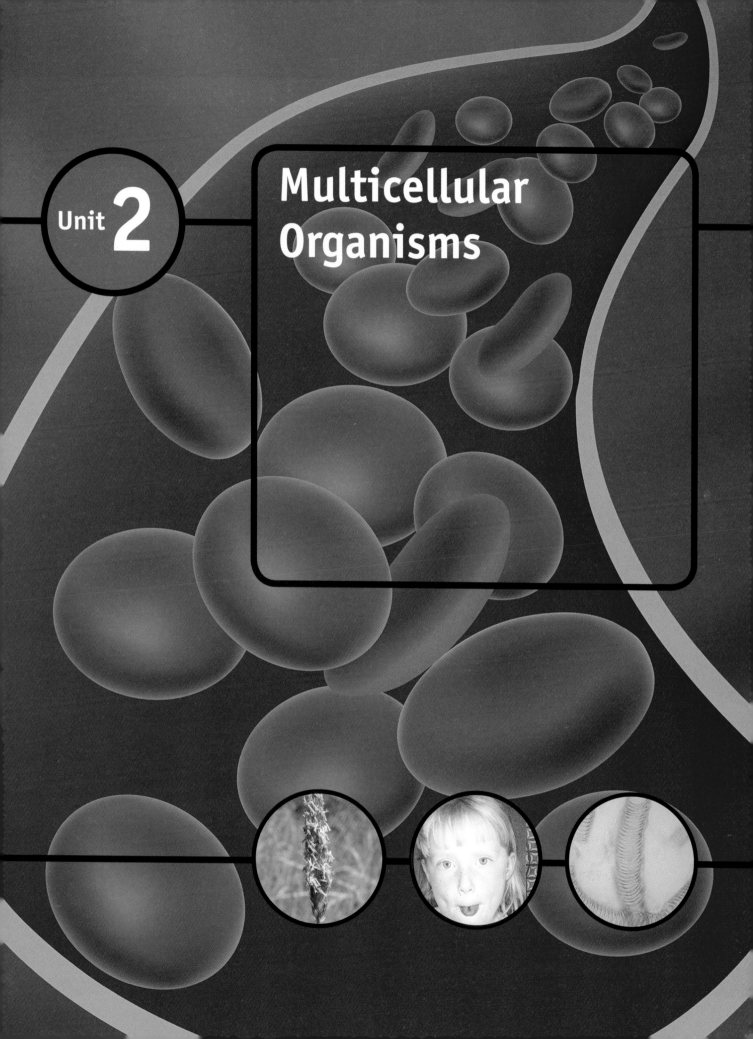

Unit 2

Multicellular Organisms

9 Cells, tissues and organs

1

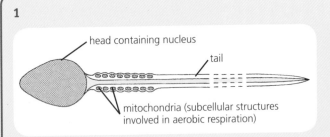

head containing nucleus

tail

mitochondria (subcellular structures involved in aerobic respiration)

Figure 9.1

a) Identify the type of sex cell shown in Figure 9.1. (1)

b) Give TWO reasons why this cell is well suited to its function. (2)

2 Figure 9.2 shows a transverse section of a young root.

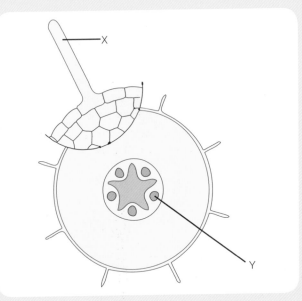

X

Y

Figure 9.2

a) Identify structure X and state its function. (2)

b) Name TWO types of cell that would be present in tissue Y and compare them with respect to their structure and function. (4)

c) i) Does a division of labour exist in a root?

ii) Explain your answer. (2)

3 Figure 9.3 shows the epithelial tissue that makes up the surface layer of a villus in the human small intestine. State TWO ways in which this tissue:

a) resembles (2)

b) differs from

the epithelium that lines the trachea. (2)

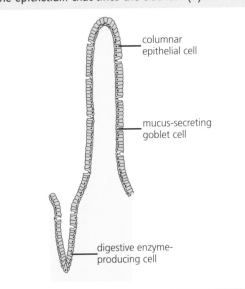

columnar epithelial cell

mucus-secreting goblet cell

digestive enzyme-producing cell

Figure 9.3

4 a) Figure 9.4 shows a multicellular plant. Which row in Table 9.1 correctly matches organs 1, 2 and 3 with their functions? (1)

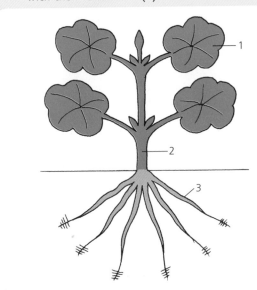

1

2

3

Figure 9.4

	Organ		
	1	**2**	**3**
A	Support	Photosynthesis	Anchorage
B	Photosynthesis	Support	Anchorage
C	Photosynthesis	Anchorage	Support
D	Anchorage	Support	Photosynthesis

Table 9.1

b) Figure 9.5 shows a different type of multicellular plant. Which row in Table 9.2 correctly matches organs 4, 5 and 6 with their functions? (1)

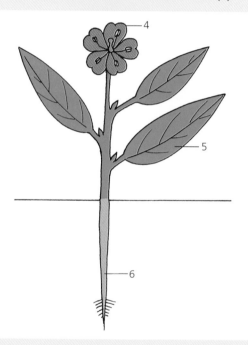

Figure 9.5

	Organ		
	1	**2**	**3**
A	Reproduction	Water absorption	Gaseous exchange
B	Gaseous exchange	Reproduction	Water absorption
C	Water absorption	Gaseous exchange	Reproduction
D	Reproduction	Gaseous exchange	Water absorption

Table 9.2

5 Read the passage and answer the questions that follow it.

If the nucleus of a cell or some other structure such as a chloroplast is removed from a cell, the isolated part cannot survive on its own. A complete cell is the smallest unit that can lead an independent life. This is neatly illustrated by unicellular *Amoeba*, a tiny animal which shows all the characteristics of living things despite the fact that it consists of only one cell.

Multicellular animals and plants are made of more than one cell. A human adult's body is composed of approximately 60 trillion cells. Instead of each one of these cells performing every function vital for life, it is more efficient for certain cells to become specialised to do particular jobs. A red blood cell, for example, is shaped like a biconcave disc, as shown in Figure 9.6. This shape presents a large surface area through which oxygen can pass into the cell before being transported to all parts of the body. As a result of its shape, a red blood cell is very good at its job.

A group of similar cells working together and carrying out a particular function is called a tissue. For example, muscle tissue brings about movement and nerve tissue carries messages. A group of different tissues, in turn, make up an organ. For example, the heart is made of different tissues such as muscle, connective and nerve tissues working together to pump blood round the human body.

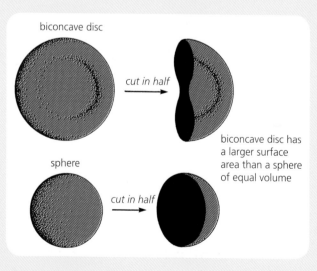

Figure 9.6

a) Explain why a cell is often described as 'the basic unit of life'. (1)

b) The first paragraph refers to 'all the characteristics of living things'. Identify FOUR of these. (4)

c) Distinguish between unicellular and multicellular living things. (1)

d) Another way of expressing the number 1000 is 10^3.

   i) A human baby is made up of about 2 trillion cells. Express this number in the same way (where 1 trillion = 1 million million).

   ii) By how many times is the number of cells in a human adult greater than that in a baby? (2)

e) Why is a red blood cell a good example of a cell whose structure is ideally suited to the job that it does? (2)

f) What general name is given to a group of similar cells working together to do a particular job? (1)

g) Human skin is composed of many structures such as blood vessels, epidermal cells, nerve endings and sweat glands.

   i) Is skin a tissue or an organ?

   ii) Explain your answer. (2)

h) Arrange the following in order of increasing complexity: *organ, cell, human body, tissue*. (1)

6 Figure 9.7 shows levels of organisation in the human body. Match numbers 1–7 with the following terms: *cell, molecule, organ, organelle, organism, system, tissue*. (6)

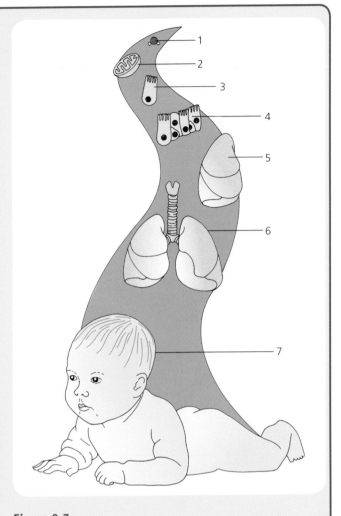

Figure 9.7

10   Stem cells and meristems

1 The following list gives the steps involved in a stem-cell-based skin graft.

   A   stem cells are sprayed over the area damaged by burning

   B   stem cells present in the skin sample are loosened and isolated using enzymes

   C   missing skin is regenerated from the stem cells

   D   stem cells are cultured to form a suspension

   E   a small sample of skin is taken from an area close to the site of injury

     a) Arrange the stages into the correct order. (1)

     b) Suggest why the skin sample is taken from an area as close as possible to the site of injury. (1)

     c) Give TWO reasons why this type of skin graft is preferable to the traditional method. (2)

     d) Briefly explain why there is no danger of tissue rejection occurring following this type of skin graft. (1)

2 The spinal cord is protected by a column of backbones called vertebrae which are grouped together into distinct regions as shown in Figure 10.1. The data in Table 10.1 refer to the results of a 5-year programme of treatment during which people who had suffered spinal cord injuries were given injections of stem cells.

   a) Copy and complete the two blank columns in the table. (2)

   b) Present the data you have calculated as a two-tone bar chart. (4)

   c) Among the people suffering damage to the cervical region of the spinal cord, by how many times did those showing an improvement outnumber those showing no improvement? (1)

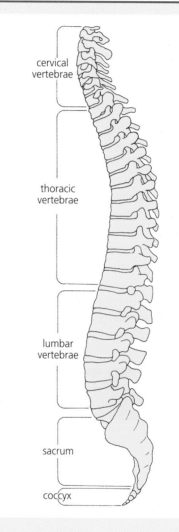

Figure 10.1

Region of spinal cord injured	Number of cases	Number of cases showing improvement	Percentage number of cases showing improvement	Number of cases not showing improvement	Percentage number of cases not showing improvement
Cervical	220	165		55	
Thoracic	180	108		72	
Lumbar	300	159		141	

Table 10.1

d) What relationship seems to exist between percentage number of cases showing improvement and distance of the injured site from the brain? (1)

e) It could be argued that the patients who showed improvement would have done so whether or not they had been treated with stem cells. What control would need to be run to counter this argument? (1)

3 The data in Table 10.2 refer to the results of an investigation into the effect of two chemicals A and B on cultures of stem cells.

a) Supply the data that should have been present in boxes X and Y. (2)

b) Plot a line graph of the three sets of mean values on the same sheet of graph paper. (5)

c) Draw TWO conclusions from the investigation. (2)

d) What is the purpose of including treatment C in the investigation? (1)

e) Why was each treatment done in triplicate? (1)

f) Identify THREE environmental factors that must be kept constant during this investigation. (3)

g) i) Calculate the percentage increase in cell number in flask 1 of treatment A after 10 days.

ii) Calculate the percentage decrease in cell number in flask 1 of treatment B after 10 days. (2)

4 Figure 10.2 shows the early development of a side root in a plant.

a) Match letters A–E with the following answers: *epidermal cell, meristematic cell, phloem tissue, root cap, xylem vessel* (4)

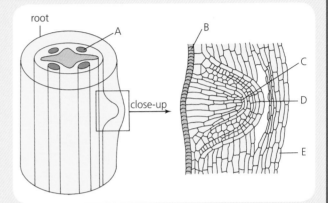

Figure 10.2

b) i) Identify the TWO letters that each represent an example of a specialised cell.

ii) Which letter represents an unspecialised cell? (3)

Treatment	Details of treatment	Flask	Number of viable stem cells $\times 10^5$ per cm^3					
			At start	Day 2	Day 4	Day 6	Day 8	Day 10
A	Growth medium + chemical A	1	5.0	5.9	7.6	10.2	14.2	16.5
		2	5.0	6.2	7.9	9.7	14.4	16.9
		3	5.0	6.2	7.3	10.1	14.3	16.1
		Mean	**5.0**	**6.1**	**7.6**	**10.0**	**14.3**	**[X]**
B	Growth medium + chemical B	1	5.0	4.2	3.6	2.4	1.8	0.5
		2	5.0	4.3	3.6	2.8	1.6	0.5
		3	5.0	4.4	3.3	2.6	1.7	0.5
		Mean	**5.0**	**4.3**	**3.5**	**2.6**	**1.7**	**0.5**
C	Growth medium only	1	5.0	5.7	6.1	6.1	6.3	6.4
		2	5.0	5.9	6.0	6.2	6.3	6.3
		3	5.0	**[Y]**	6.2	6.3	6.3	6.2
		Mean	**5.0**	**5.7**	**6.1**	**6.2**	**6.3**	**6.3**

Table 10.2

5 Figure 10.3 shows a section through a shoot tip meristem. Which of the diagrams in Figure 10.4 shows the correct appearance of this shoot tip at the formation of the next new leaf? (1)

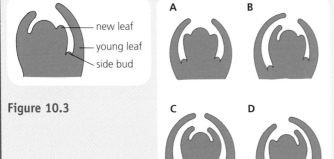

Figure 10.3

Figure 10.4

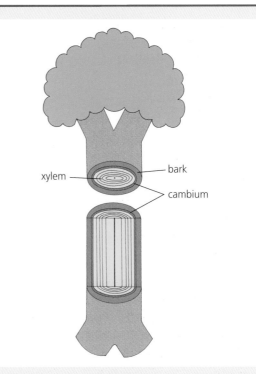

6 Figure 10.5 shows a twig from a tree in winter.

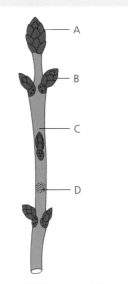

Figure 10.5

a) Which structure, by its growth the next year, will increase the length of this twig? (1)

b) Which structure may grow into a new branch? (1)

7 Some plants, such as trees, continue to grow and increase in size year after year. Therefore they need to develop additional xylem cells (vessels) every year. This new tissue is produced during spring, summer and autumn by a meristem called cambium. This meristem takes the form of a continuous cylinder inside the plant, as shown in Figure 10.6. Each year it forms woody xylem on its inside.

Figure 10.6

a) Give TWO reasons why a plant that increases in size year after year needs additional xylem vessels. (2)

b) i) At which time of the year does cambium not produce new xylem?

   ii) Suggest why. (2)

c) i) How old is the tree in Figure 10.6?

   ii) Explain how you arrived at your answer. (2)

d) Which one of the following sets of conditions during the summer months would result in the poorest production of xylem by cambium? (1)

   A warm weather, infestation by greenfly and lack of rainfall

   B cold weather, infestation by greenfly and abundant sunshine

   C warm weather, lack of rainfall and abundant sunshine

   D cold weather, lack of rainfall and infestation by greenfly

8 Read the passage and answer the questions that follow it.

Many flowering plants have extensive powers of regeneration – the ability to replace lost or damaged parts. A piece of meristem cut from a root or shoot tip will often develop roots and shoots and eventually regenerate the entire plant. This phenomenon is put to good use in a plant-cloning laboratory where thousands of genetically identical copies of a commercially desirable plant are produced by micropropagation.

One technique involves cutting out tiny pieces of meristem from the buds of a parent plant and growing them on sterile nutrient agar in culture tubes. Each piece of tissue is kept in optimum conditions of light and temperature until it develops into a miniature plant. It is then checked carefully for disease before being supplied in a batch to a commercial grower who wishes to obtain a uniform crop such as fruit that all ripen at the same time.

This seemingly ideal situation does, however, carry an element of risk. If a new disease-causing organism arrives that can attack the propagated plants, then the whole crop may be lost.

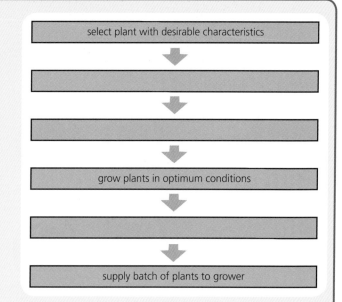

```
┌─────────────────────────────────────────────┐
│  select plant with desirable characteristics │
└─────────────────────────────────────────────┘
                       ▼
┌─────────────────────────────────────────────┐
│                                             │
└─────────────────────────────────────────────┘
                       ▼
┌─────────────────────────────────────────────┐
│                                             │
└─────────────────────────────────────────────┘
                       ▼
┌─────────────────────────────────────────────┐
│        grow plants in optimum conditions     │
└─────────────────────────────────────────────┘
                       ▼
┌─────────────────────────────────────────────┐
│                                             │
└─────────────────────────────────────────────┘
                       ▼
┌─────────────────────────────────────────────┐
│         supply batch of plants to grower     │
└─────────────────────────────────────────────┘
```

Figure 10.7

a) The flow diagram in Figure 10.7 is intended to give a summary of the procedure described in the passage. Copy it and complete the blanks. (3)

b) Name TWO environmental conditions given in the passage that would be kept at optimum levels to promote plant growth. (2)

c) State ONE precaution that is taken to prevent the young plants from becoming diseased. (1)

d) One of the benefits gained by the commercial grower is that there is no need to return again and again to pick the fruit crop. Identify the phrase that tells you why one visit to the crop at harvest time will be sufficient. (1)

e) What term given in the passage means 'producing a population of identical plants'? (1)

f) i) What could cause a grower to lose the entire crop?
   ii) Explain why. (2)

11 Control and communication

1 Figure 11.1 shows a section of the human brain.

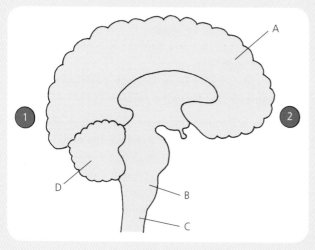

Figure 11.1

a) Would the person's face be at side 1 or side 2? (1)
b) Name parts A, B, C and D. (4)
c) Decide whether each of the following statements about the brain is true or false and use T or F to indicate your choice. Where a statement is false, give the word that should have been used in place of the one in coloured print. (6)
   i) The **medulla** brings about an increase in the rate of breathing after vigorous exercise.
   ii) The **cerebrum** is responsible for creativity and personality.
   iii) The **cerebellum** contains the long-term memory.
   iv) The **medulla** receives information from the semi-circular canals.
   v) The **cerebellum** is needed for effective muscular coordination.
   vi) The **cerebrum** makes the heart beat more rapidly during exercise.

2 The sensitivity of a boy's skin to temperature was investigated as follows. A grid of 2 mm squares was printed on the palm of his right hand and he was blindfolded. The experimenter then tested each square for its sensitivity to both heat and cold in a random manner planned in advance. To do this she used a blunt pin probe (see Figure 11.2) which had been

placed briefly in either very hot water or icy water. Figure 11.3 shows the results.

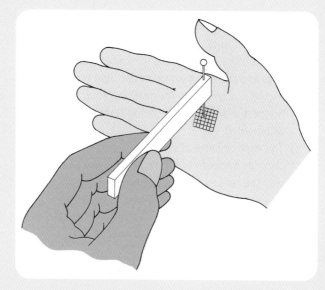

Figure 11.2

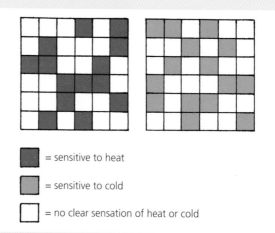

■ = sensitive to heat

■ = sensitive to cold

□ = no clear sensation of heat or cold

Figure 11.3

a) How many squares communicated information to the brain that resulted in:
   i) the sensation of heat only?
   ii) the sensation of cold only?
   iii) the sensation of both heat and cold?
   iv) no clear sensation of either heat or cold? (4)

b) Which of the following hypotheses are supported by the results of this experiment? (2)

   1 Temperature receptors exist in a continuous, unbroken layer under the surface of the skin.

   2 Temperature receptors in the skin are distributed as tiny temperature-sensitive 'spots' with spaces between them.

   3 Every temperature receptor is sensitive to both heat and cold.

   4 The skin contains two types of temperature receptor, some that detect heat and some that detect cold.

c) Why did the experimenter test the squares in the grid in a *random* manner rather than testing them methodically in turn for heat and cold? (1)

d) Suggest why the person being tested was blindfolded. (1)

e) The experimenter allowed intervals of a few seconds between applications of the probe. What was the reason for this procedure? (1)

f) In what way would the experiment need to be expanded so that a generalisation could be made about

   i) the sensitivity of this person's skin to temperature

   ii) the sensitivity of human skin in general to temperature? (2)

g) How could the experiment be adapted to investigate the distribution of pain receptors in the skin? (1)

h) Explain why pain such as toothache is of benefit to the human body. (1)

3 Table 11.1 refers to the rate of conduction of nerve impulses. Some invertebrate animals have giant nerve fibres in their bodies in addition to normal nerve fibres.

a) Draw TWO conclusions about the rate of conduction of nerve impulses from the data in the table. (2)

b) Suggest why it is of advantage to a squid to have giant nerve fibres running along the length of its muscular coat. (1)

c) By how many times, on average, is the rate of conduction of an impulse in a nerve of a mammal faster than that in a normal nerve fibre of a cockroach? (1)

4 Copy and complete Table 11.2 using information contained in the following passage. (11)

A reflex action is a simple act of behaviour whose function is protective. For example, the eyelid muscle contracts when an object touches the eye. This results in blinking, which helps to prevent damage to the eye. Similarly, when foreign particles such as pepper enter the nasal tract, a sudden contraction of the chest muscles makes the person sneeze and remove unwanted particles from the nose.

Although reflex actions are involuntary, some can be partly altered by voluntary means; the person can, to a certain extent, resist blinking or sneezing. Some reflex actions cannot, however, be altered by voluntary means. When food is present in the gut, muscles in the gut wall contract bringing about peristalsis. This muscular movement also ensures efficient digestion by mixing the food thoroughly with digestive enzymes.

Animal		'Warm-blooded' (W) or 'cold-blooded' (C)	Rate of conduction of nerve impulse (m/s)
Snake		C	10–30
Mammal		W	30–120
Frog		C	7–30
Bird		W	30–120
Squid	normal fibre	C	4.3
	giant fibre		18–30
Earthworm	normal fibre	C	0.6
	giant fibre		10–30
Cockroach	normal fibre	C	1.5–6.0
	giant fibre		9–12

Table 11.1

Reflex action	Stimulus	Response	Protective function	Can be altered partly by voluntary means?
Blinking		Contraction of eyelid muscle		Yes
	Presence of food in gut		Ensures movement and therefore efficient digestion of food	No
	Foreign particles in nasal tract	Sudden contraction of chest muscles		
Dilation of pupil				

Table 11.2

In dim light, the pupil of the eye becomes dilated (enlarged) as a result of movement of the iris muscle. This reflex action improves the person's vision in poor lighting, thereby reducing the chance of an accident. Peristalsis and pupil dilation cannot be resisted or prevented by voluntary means.

5 Table 11.3 refers to the hormones that control the menstrual cycle in human females. A gonadotrophic hormone stimulates reproductive organs and an ovarian hormone is made by ovary tissue.

a) With reference to Table 11.3, identify TWO gonadotrophic hormones present in the bodies of human females. (2)

b) With reference to the table, give TWO examples of ovarian hormones. (2)

c) i) Match the four hormones with black arrows 1, 2, 3 and 4 in Figure 11.4.

   ii) Which of these hormones stimulates ovulation? (5)

Endocrine gland	Hormone	Effect
Pituitary	Follicle-stimulating hormone (FSH)	• Stimulates development of follicle in ovary • Stimulates ovary tissue to produce oestrogen
Ovary	Oestrogen	• Repairs uterus lining following menstruation • Stimulates pituitary to make luteinising hormone
Pituitary	Luteinising hormone (LH)	• Brings about ovulation • Causes follicle to become corpus luteum which secretes progesterone
Ovary (corpus luteum)	Progesterone	• Promotes thickening of uterus lining

Table 11.3

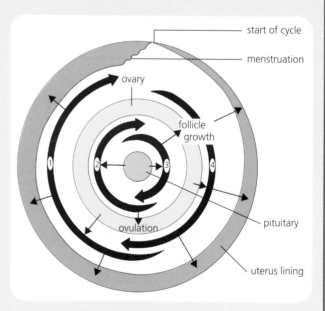

Figure 11.4

6 Table 11.4 shows the results of a diabetes survey.

Age (years)	Percentage of total number of people registered with diabetes	
	Male	Female
15–44	5.8	4.8
45–64	21.0	17.2
65–84	25.0	20.5
85 and over	2.7	2.2

Table 11.4

a) Draw TWO conclusions from the data in the table. (2)
b) Why do the percentage data in the table not add up to 100%? (1)

7 Graphs 1 and 2 in Figure 11.5 show the results of two patients, X and Y, each ingesting 100 g of glucose.
a) What was the concentration of insulin in person X's blood after 1 hour? (1)
b) What was the highest concentration of glucose reached in the blood of
    i)   person X
    ii) person Y? (2)
c) i)   How long did it take for person X's blood glucose to return to normal after ingestion of 100 g of glucose?
    ii) Where in the body would the ingested glucose be most likely to end up if it was surplus to the body's immediate requirements? (2)
d) With reference to graphs 1 and 2, give TWO pieces of evidence that support the theory that person Y has untreated diabetes type 1. (2)

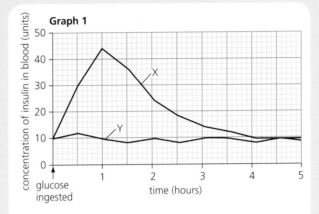

Graph 1

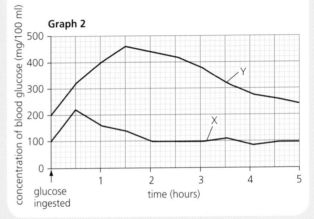

Graph 2

Figure 11.5

1 a) An average human sperm is 60 micrometres in length. Express this as a decimal fraction of a millimetre. (1 millimetre = 1000 micrometres) (1)

b) Sperm count is measured as millions of sperm per millilitre (ml) of semen. A man is considered to be fertile if his sperm count is 20 million or more. Table 12.1 shows the data for three patients at a fertility clinic.

Patient	Average volume of semen released (ml)	Average total number of sperm in semen (millions)
X	3.0	63
Y	4.0	76
Z	4.5	117

Table 12.1

i) Calculate the sperm count for patients X, Y and Z.
ii) Which of these men is most likely to be infertile? (4)

c) Suggest why mammals produce many more sperm than eggs. (1)

2 Copy and complete Table 12.2 by referring to Figure 12.1, which shows the reproductive organs of a female rat. (7)

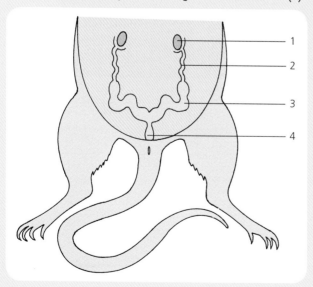

Figure 12.1

Site of:	Number of structure	Name of structure
Copulation		
Egg production		
Fertilisation		
Embryo development		

Table 12.2

3 The bar chart in Figure 12.2 shows the mean number of eggs produced by different species of bird.

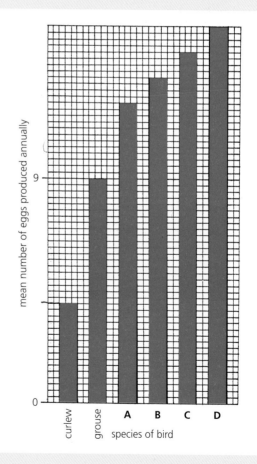

Figure 12.2

a) What is the mean number of eggs produced annually by a curlew? (1)

b) Which bar in the graph represents the robin which produces, on average, 13 eggs per year? (1)

4 Female desert locusts lay their eggs in moist sand. Table 12.3 shows the results of an investigation using ten cages. Water was added to the sand in each cage as indicated. Ten females were then released into each cage and allowed to lay egg pods (collections of eggs stuck together).

Cage	Volume of water added per 100 g sand (cm³)	Number of egg pods laid
1	0	0
2	3	2
3	6	4
4	9	13
5	12	20
6	15	24
7	18	26
8	21	20
9	24	6
10	27	0

Table 12.3

a) Draw a line graph of the data by plotting points and joining them with a ruler. (3)

b) i) What is the variable factor investigated in this experiment?

   ii) What is the optimum value of the variable factor? (2)

c) i) Work out from your graph how many egg pods would be laid if 8 cm³ of water per 100 g of sand were added.

   ii) Two different volumes of water added to 100 g of sand would each result in 22 egg pods being laid. Identify these TWO volumes of water from your graph. (3)

d) Why were as many as ten locusts used per cage? (1)

e) Suggest TWO reasons why female locusts lay their egg pods buried in sand rather than on the surface. (2)

5 Table 12.4 refers to the survival of eggs and young of several types of vertebrate animal.

a) Calculate the values omitted from boxes X, Y and Z in the table. (3)

b) i) What relationship exists between the number of eggs produced annually and the length of parental care given to the young?

   ii) What relationship exists between percentage survival of young and length of parental care? (2)

c) Suggest why some of the information in the table can only be given as approximate values. (1)

Animal	Class of vertebrate	Number of eggs produced annually	Number surviving after 1 year	Percentage survival of young	Length of time spent by parents feeding and protecting young (weeks)
Cod	Fish	4 000 000	[X]	0.01	0
Trout	Fish	3000	150	[Y]	0
Sparrow	Bird	15	12	80	2
Ptarmigan	Bird	[Z]	9	90	3
Fox	Mammal	4	4	100	10
Whale	Mammal	1	1	100	80

Table 12.4

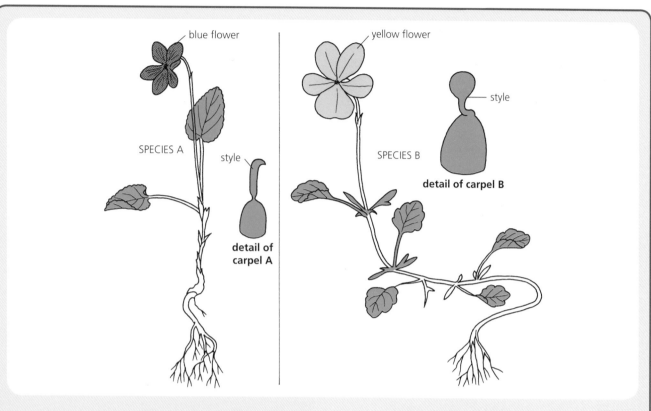

Figure 12.3

6 a) Use the following key to identify the two species of *Viola* shown in Figure 12.3. (2)

1 style expanded into ball-like stigma ..**2**
style not expanded into ball-like stigma **3**

2 long creeping horizontal stem present....................................... *Viola lutea*
no long creeping horizontal stem present .... *Viola tricolor*

3 style extended into hook-like stigma... **4**
style not extended into hook-like stigma... *Viola palustris*

4 leaf stalks (petioles) hairy **5**
leaf stalks (petioles) not hairy **6**

5 flowers sweetly scented.................. *Viola odorata*
flowers not sweetly scented .... *Viola hirta*

6 petals blue.................................... *Viola canina*
petals white .... *Viola stagnina*

b) Describe *Viola odorata*. (3)

7 Figure 12.4 shows part of a flower.

a) Which one of the following routes represents pollination? (1)

   A From 1 to 2.
   B From 2 to 3.
   C From 4 to 5.
   D From 1 to 6.

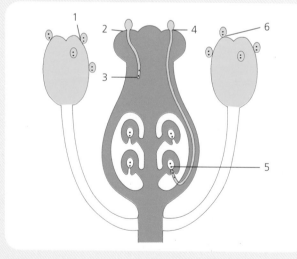

Figure 12.4

b) At which of the following points does fertilisation occur? (1)

   A 1
   B 2
   C 3
   D 5

8 Figure 12.5 shows an experiment set up to investigate the effect of two concentrations of two types of sugar on pollen tube formation by pollen grains from two types of plant – *Narcissus* (N) and *Impatiens* (I).
  a) Which TWO numbered parts of the experiment should be compared to find out the effect of concentration of:
    i) glucose on pollen tube formation in *Narcissus*?
    ii) sucrose on pollen tube formation in *Impatiens*? (2)
  b) Which of the following should be compared to find out the effect of type of sugar on pollen tube formation in *Impatiens*? (1)

A 1 and 6
B 2 and 5
C 3 and 8
D 4 and 8

  c) By which variable factor do set-ups 5 and 6 differ? (1)
  d) Explain why a comparison of set-ups 3 and 6 is invalid. (1)
  e) Identify TWO environmental factors that must be kept constant in all eight parts of the experiment to make it valid. (2)
  f) Suggest what could be done to improve the reliability of the results? (1)

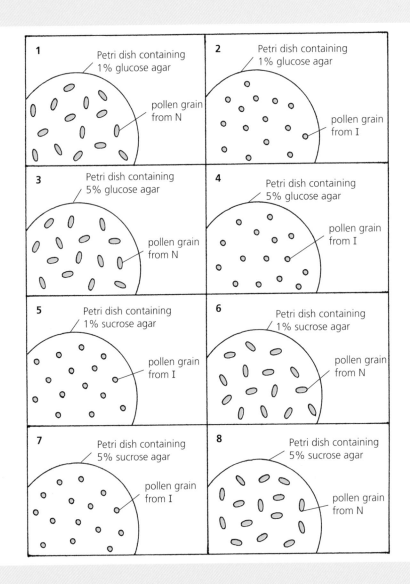

Figure 12.5

9 Some flowers are pollinated by insects and others by wind, as shown in Figure 12.6. Table 12.5 compares the two types of pollination. Copy and complete the table using the boxed answers given in Figure 12.7. (7)

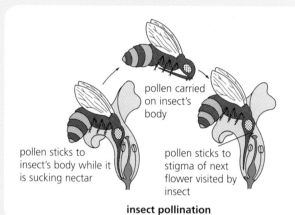

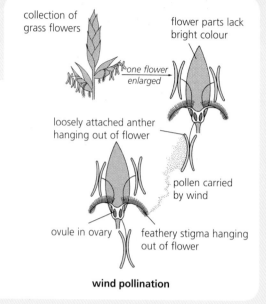

Figure 12.6

Insect-pollinated flower		Wind-pollinated flower	
Structural feature	**Reason**	**Structural feature**	**Reason**
	To attract insects which will eat nectar and collect pollen		No visit from insect required
	To be in a position where insects are likely to brush against them	Anthers loosely attached and hanging out of flower	
Pollen grains sticky or with rough surface		Pollen grains light and smooth	
Stigmas inside the flower with sticky surface		Stigmas hanging out of flower and feathery in structure	

Table 12.5

To enable them to be shaken and to allow pollen to be carried away by wind	To be in a good position and present a large surface area for trapping pollen	Flower large with brightly coloured petals, scent and nectar	Anthers firmly attached inside flower
To enable them to be carried in air currents without sticking together	To be in a position where insect is likely to brush against them and leave pollen stuck to them	To enable them to stick to insect's body easily	Flower small, lacking bright colour, scent and nectar

Figure 12.7

13 Variation and inheritance

1 The histogram in Figure 13.1 shows the variation in mass of the individual grapes that made up a bunch.
   a) Describe the distribution of grape mass in terms of:
      i) range
      ii) most common value for this population. (2)
   b) How many grapes were found to have the most common value? (1)
   c) What was the least common mass in this survey? (1)
   d) How many grapes had a mass of 2.8 g? (1)
   e) How many grapes were weighed? (1)
   f) What percentage of grapes had a mass of 2.5 g? (1)
   g) How many grapes weighed more than 2.7 g? (1)
   h) Calculate the average (mean) mass of a grape for this population. (1)

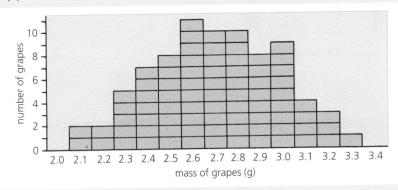

Figure 13.1

2 Fifty pupils had their left index finger measured. The lengths are listed in Table 13.1.

Pupil number	Length of left index finger (mm)	Pupil number	Length of left index finger (mm)	Pupil number	Length of left index finger (mm)	Pupil number	Length of left index finger (mm)
1.	48	13	62	25	67	38	74
2	51	14	62	26	68	39	75
3	52	15	62	27	68	40	75
4	53	16	63	28	69	41	76
5	55	17	63	29	70	42	78
6	56	18	65	30	70	43	78
7	57	19	65	31	71	44	79
8	58	20	65	32	71	45	81
9	58	21	66	33	71	46	83
10	60	22	66	34	72	47	84
11	60	23	66	35	72	48	85
12	61	24	67	36	73	49	88
				37	74	50	91

Table 13.1

a) Does length of index finger show discrete or continuous variation? (1)

b) Divide the range into ten subsets of 5 mm (with the first subset being 45–49 mm) and then present the information as a histogram. (4)

c) How many pupils have a left index finger in the subset 70–74 mm? (1)

d) i) What is the most common subset of the range of index finger length?

ii) How many pupils have a left index finger that belongs in this subset? (2)

e) How many pupils have a left index finger of 75 mm or more? (1)

3 In fruit flies, straight wing (S) is dominant to curled wing (s). Figure 13.2 shows the results of an investigation carried out to examine the phenotypes arising from a single gene cross involving the wing-type gene.

	straight wing	curled wing
parents of F₁	6 true-breeding males	6 true-breeding females
F₁	168 flies of both sexes	0
parents of F₂	6 males from F₁	6 true-breeding females
F₂	81 flies of both sexes	87 flies of both sexes

Figure 13.2

a) Present the information in the standard diagrammatic form including a Punnett square in your answer. (4)

b) Explain why no curled-winged flies were produced in the F₁ generation? (1)

c) i) What is the expected ratio of straight to curled in the F₂ generation?

ii) What is the expected number of straight to curled in the F₂ generation?

iii) Why do the actual results vary slightly from the expected ones? (4)

4 In tobacco plants the gene for leaf colour has two forms (alleles), green and white, where green (G) is dominant to white (g). A heterozygous green plant was crossed with a white plant.

a) Give the genotypes of these two parent plants. (1)

b) State:
i) the genotypes of the offspring that would be produced
ii) the ratio in which they would be expected to occur. (2)

c) What would be the genotype of a homozygous green plant? (1)

5 Figure 13.3 shows a family tree that refers to tongue rolling. Figure 13.4 shows a family tree that refers to hair type. In humans, the allele for tongue rolling ability (R) is dominant to that for inability to roll the tongue (r) and the allele for wavy hair (H) is dominant to the allele for straight hair (h).

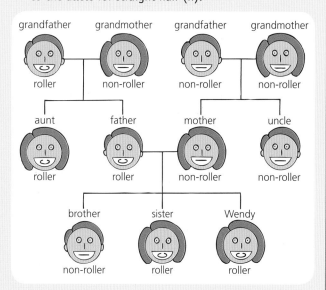

Figure 13.3

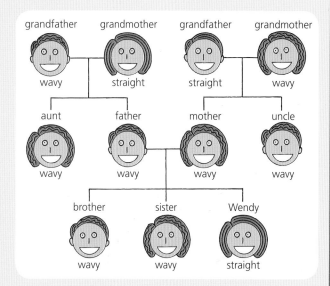

Figure 13.4

a) From which parent did Wendy inherit her tongue-rolling ability? (1)

b) From which of her grandparents did Wendy receive the genetic information that gave her straight hair? (2)

c) Using a square to represent a male and a circle to represent a female, copy each family tree and write in the genotypes. (8)

6 In humans, a patch of white hair at the front of the head is called a white forelock. Possession of a white forelock (F) is dominant to lack of a white forelock (f). Read the details that follow about a family.

Mandy's mother and maternal grandparents are all homozygous for a white forelock. Although Mandy's father lacks a white forelock, both of his parents have a white forelock.

a) Draw a diagram to show Mandy's family tree and include a key to explain the meanings of the symbols you have chosen. (4)

b) Add the genotypes to the family tree. (4)

c) What is Mandy's phenotype with respect to this inherited characteristic? (1)

7 Cystic fibrosis is an inherited disorder of the human body. It affects mucus production, causing blockage of tiny air passages in the lungs. It is due to a recessive gene form.

a) Copy and complete the family trees shown in Figure 13.5 by inserting the missing genotypes. (4)

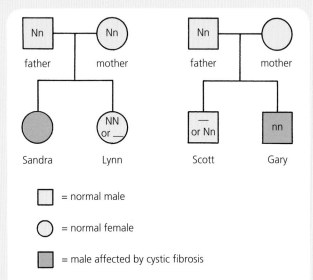

Figure 13.5

b) Lynn and Scott are intending to marry so they seek the advice of a genetic counsellor. The counsellor studies their family trees and works out that if the couple marry then this would result in one of the following crosses:

A NN × NN

B NN × Nn

C Nn × Nn

i) Which TWO crosses involve no risk of producing affected children?

ii) In the remaining cross, what is the chance of each child being affected?

iii) Using a diagram that includes a Punnett square, show how you arrived at your answer to i). (5)

8 Grain colour in wheat is an example of polygenic inheritance. It can be explained on the basis that it is controlled by three genes whose dominant alleles code for red pigment and have an additive effect.

The genotypes of plants that make grain with 100% red pigment and those that make grain with 0% pigmentation (white) are represented in Figure 13.6, which also shows the outcome of crossing these two true-breeding plants. Figure 13.7 shows a Punnett square which represents selfing the F_1 generation. It gives the eight types of pollen and ovules that would be involved.

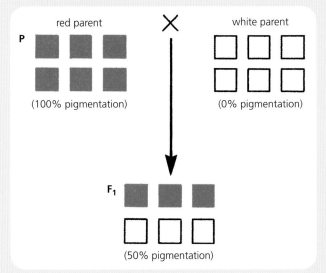

Figure 13.6

a) Copy the Punnett square in Figure 13.7 and complete it by inserting the number of red alleles that would be present in each zygote following fertilisation. (A few have been done for you.) (2)

b) i) Assuming that each red allele contributes one unit of red pigment to the wheat grain's phenotype, state how many shades of red would be present in the F_2 generation.

ii) Which shade of red is most common?

iii) Which shade of red would be least common?

iv) What percentage of grains in an F_2 generation would be expected to have the shade of red determined by one red allele? (4)

c) From the data, express as a ratio the number of F_2 grains whose red colour is determined by three alleles to those whose red colour is controlled by two alleles. (1)

Figure 13.7

14   Transport systems in plants

1 Figure 14.1 shows a transverse section of a plant's stem and close-ups of three of its tissues.

a) Which of the four answers **A–D** given on the right correctly describes tissue Q? (1)

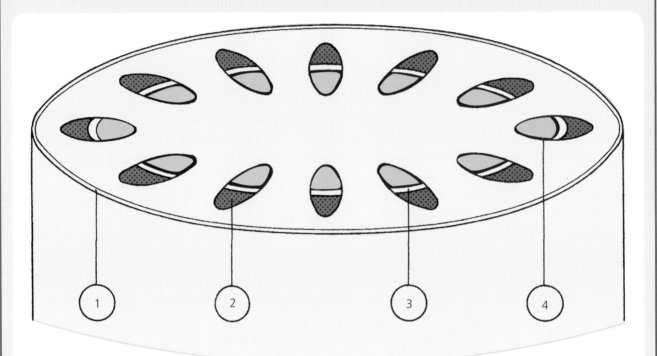

tissue Q   tissue R   tissue S

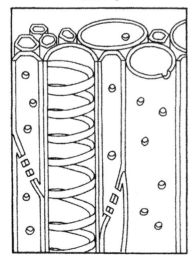
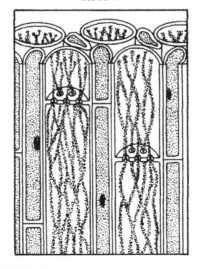
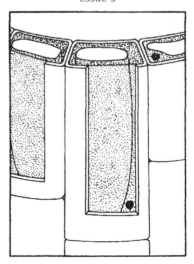

Figure 14.1

b) Which of the four answers correctly describes tissue R? (1)

c) Which of the four answers correctly describes tissue S? (1)

Answers:

A It is found at location 1 and it protects the plant.

B It is found at location 2 and it is the site of sugar transport.

C It is found at location 3 and it supports the plant.

D It is found at location 4 and it is the site of water transport.

d) What is the name of tissue R? (1)

  A cortex
  B xylem
  C phloem
  D epidermis

2 In the transverse section of part of a green leaf shown in Figure 14.2, which arrow shows the correct direction of water flow? (1)

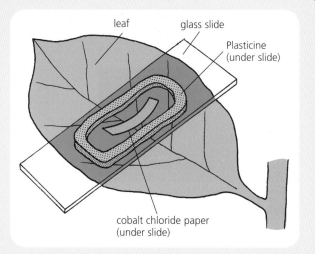

Figure 14.3

b) i) Identify a possible source of experimental error in the above experiment.

  ii) Suggest a method of minimising the source of error you identified. (2)

c) It could be argued that the cobalt chloride paper was going to change colour whether or not it had been in contact with the leaf. Describe fully the control that should have been included in the experiment. (2)

4 The electronic balance in Figure 14.4 shows the final reading of a weight potometer which has been allowed

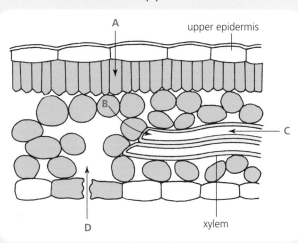

Figure 14.2

3 Cobalt chloride paper is blue when dry and pink when dampened by water vapour. In the experiment shown in Figure 14.3, the strip of paper was blue when it was taken out of the desiccator (drying chamber). The paper was taken to a plant on a window sill at the other side of the laboratory and placed in the enclosed area of leaf surface as shown. After ten minutes in contact with the leaf, the paper had turned pink. It was therefore concluded that ten minutes is the time needed for the leaf surface to give out enough water vapour to affect the paper.

a) What colour change would be observed when moist cobalt chloride paper is placed in a desiccator for several hours? (1)

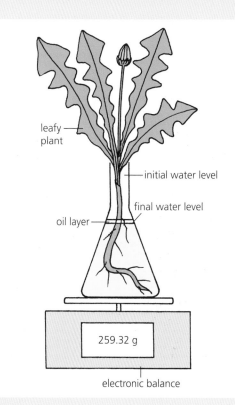

Figure 14.4

to lose water for 2 days in light. Table 14.1 records the results of the experiment and the volume of water needed after 2 days to restore the water to its initial level.

Initial mass of potometer (g)	283.80
Mass after 2 days (g)	[Q]
Mass of water lost (g)	[R]
Volume of water lost (cm³)	24.48
Volume of water needed to restore initial level (cm³)	25.00

Table 14.1

a) Supply the values missing from boxes Q and R in the table. (2)
b) Calculate the average mass of water lost per hour by the plant. (1)
c) Explain why the volume of water needed to restore the initial level differed from the volume lost by the plant. (1)
d) i) Predict the effect on water loss of keeping the weight potometer in total darkness for 2 days.
   ii) Explain your answer. (2)

5 The escape of water vapour through a leaf's stomata can be prevented by blocking the pores with Vaseline. Figure 14.5 shows an experiment involving four leaves that have been treated in different ways. Table 14.2 gives the results of the experiment after 2 days in light.

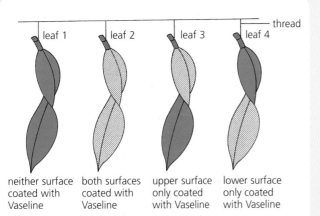

neither surface coated with Vaseline   both surfaces coated with Vaseline   upper surface only coated with Vaseline   lower surface only coated with Vaseline

Figure 14.5

a) Explain why leaf 1 lost the most mass. (1)
b) Explain why leaf 2 did not lose any mass. (1)
c) i) Calculate the values for boxes X and Y in the table.
   ii) Comparing leaves 3 and 4, which showed the higher percentage loss in mass?

Leaf	Initial mass(g)	Final mass(g)	Loss in mass(g)	Percentage loss in mass
1	10.00	8.70	1.30	13.00
2	10.13	10.13	0.00	0.00
3	9.90	8.91	[X]	10.00
4	10.20	9.89	0.31	[Y]

Table 14.2

iii) What conclusion can be drawn from the results for leaves 3 and 4 about the distribution of the stomata on this type of leaf? (4)

d) Figure 14.6 shows a second experiment using leaves from a water lily plant.

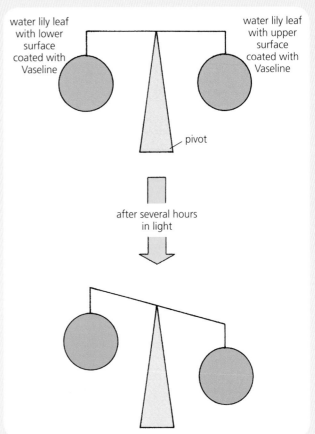

water lily leaf with lower surface coated with Vaseline

water lily leaf with upper surface coated with Vaseline

pivot

after several hours in light

Figure 14.6

i) Which leaf lost more water?
ii) What can be concluded about the location of most (or all) of the stomata in a water lily leaf?
iii) Suggest why this adaptation is of survival value to a water lily plant. (4)

6 Leafy shoots P and Q in Figure 14.7 had been growing for 2 years before they were cut and ringed.

The ringing procedure involved the removal of a section of outer tissue ('bark') from each stem so that the inner woody tissue was left intact. Wax seals were then added to the stems as shown in the diagram.

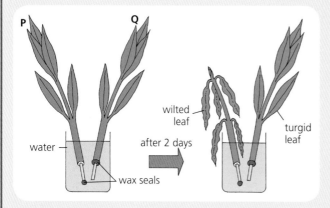

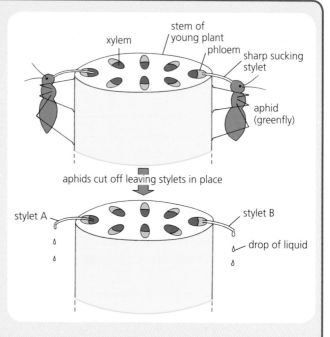

Figure 14.7

a) Which transport tissue was sealed in
   i) shoot P?
   ii) shoot Q? (2)
b) Explain fully the difference in the appearance of the leaves after 2 days. (2)

7 An aphid (greenfly) has a mouthpart called a stylet which is shaped like a syringe needle. Once settled on a plant, an aphid probes the plant's tissues using its stylet until it finds the transport tissues. Figure 14.8 shows two aphids on the stem of a young plant.
   a) i) Which substance would be escaping from the plant in stylet A?
      ii) Explain your answer. (2)
   b) i) Which substance would be escaping from the plant in stylet B?
      ii) Explain your answer. (2)

Figure 14.8

c) The region of stem above the part shown supports a flower but no leaves.
   i) In which direction would the liquid be travelling in the phloem tissue in the diagram?
   ii) Explain your answer. (2)

8 Figure 14.9 shows a close relationship between two plants, gorse and dodder.
   a) Why is dodder unable to photosynthesise? (1)
   b) Match A and B with the names of the two plants. (1)
   c) Identify tissues X and Y. (2)
   d) By what means does dodder obtain a supply of soluble sugar? (1)
   e) A mature dodder plant lacks roots. By what means does it obtain water and mineral salts? (1)
   f) Which of the plants in Figure 14.9 is the parasite and which is the host? (1)

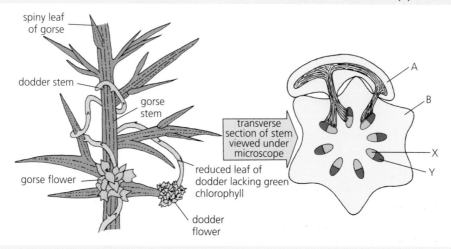

Figure 14.9

15 Animal transport and exchange systems

1 a) i) Which has the larger surface area to volume ratio, a frog or a unicellular animal?
   ii) Which of these animals needs a further oxygen-absorbing surface in addition to its body's thin outer layer?
   iii) Identify the structures that contain this oxygen-absorbing surface. (3)
  b) Name the corresponding oxygen-absorbing surface in a fish. (1)

2 Figure 15.1 shows part of the human circulatory system.

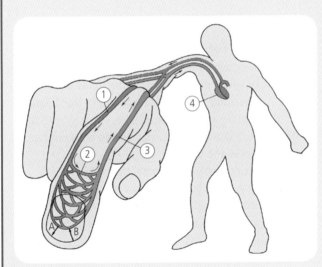

Figure 15.1

  a) i) Name structures 1–4.
     ii) Which of these contains valves? (6)
  b) i) Name TWO substances that could be passing from the blood along arrow A to the living cells below the fingernail.
     ii) Name ONE substance that could be passing along arrow B from the living cells to the blood. (3)

3 The bar chart in Figure 15.2 shows the rate of blood flow in various parts of a person's body under differing conditions of exercise.
  a) Using the information in the bar chart, copy and complete Table 15.1. (5)

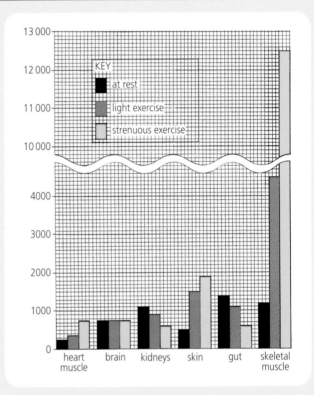

Figure 15.2

Part of body	Rate of blood flow (cm³/min)		
	At rest		Strenuous exercise
Heart muscle		350	750
	750	750	
Kidneys		900	600
	500		1900
Gut	1400	1100	
Skeletal muscle	1200		

Table 15.1

  b) Identify the labels that should have been added to the x- and y-axes on the graph in Figure 15.2. (2)

c) i) What effect does increasingly strenuous exercise have on blood flow in skeletal muscle?

ii) Suggest the reason for this. (2)

d) Which other body part(s) show the same trend in response to increase in exercise as:

i) skeletal muscle?

ii) gut? (2)

e) i) Which body part's rate of blood flow remains unaffected by exercise?

ii) Suggest why. (2)

f) i) In what way would the appearance of facial skin change as a result of strenuous exercise?

ii) Explain your answer. (2)

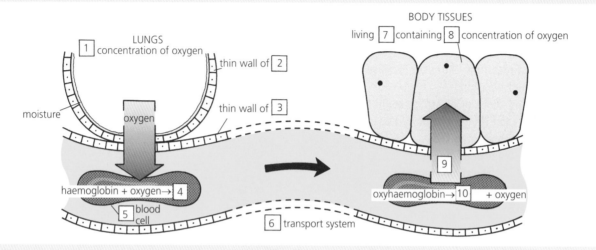

Figure 15.3

4 Figure 15.3 shows the role played by haemoglobin in the body.

a) Match boxes 1–10 with the following answers: *alveolus, blood, capillary, cell, haemoglobin, high, low, oxygen, oxyhaemoglobin, red* (9)

b) Why is the blood in the pulmonary arteries dark red and the blood in the pulmonary veins bright red? (2)

5 One cubic millimetre of a person's blood was found to contain 5 350 904 red blood cells and 8948 white blood cells. Express these figures as a whole number ratio. (1)

6 The graph in Figure 15.4 shows the relationship between breathing and the concentration of carbon dioxide in inhaled air. Choose the ONE correct answer to each of the following questions.

a) When the carbon dioxide concentration in inhaled air is increased from 3% to 6%, the average rate of breathing (in breaths/min) increases by

A 9.5

B 13.0

C 1050.0

D 1400.0 (1)

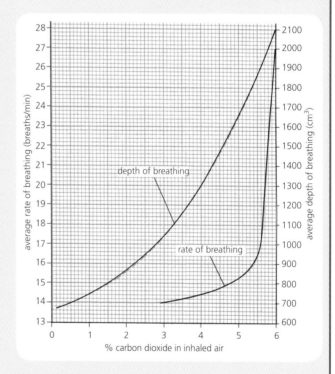

Figure 15.4

b) When the carbon dioxide concentration in inhaled air is increased from 2% to 6% the average depth of breathing (in cm^3) increases by

A 12.4

B 13.0

C 1240.0

D 1400.0 (1)

c) As the carbon dioxide concentration in inhaled air increases

A rate and depth of breathing increase at the same rate

B rate of breathing shows an increase before depth of breathing

C rate and depth of breathing begin to increase at the same concentration of CO_2

D depth of breathing shows an increase before rate of breathing. (1)

7 Figure 15.5 shows the three stages of a peristaltic wave in a small section of the gut. Part of stage B has been omitted.

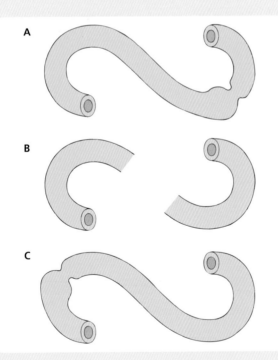

A

B

C

Figure 15.5

a) i) Copy and complete stage B of the diagram.

ii) Add the following labels to your diagram: muscular wall of alimentary canal, region where circular muscle is contracted, region where circular muscle is relaxed, food undergoing digestion.

iii) Add a large arrow to your diagram to show the direction in which the food is travelling. (6)

b) i) Name the region of the alimentary canal represented by the diagram.

ii) Justify your choice. (2)

c) Name TWO other regions of the gut where peristalsis occurs. (2)

8 Figure 15.6 shows the daily turnover of calcium in a human adult who has consumed 900 mg of calcium. Bone is continually being remodelled by the processes of bone formation and reabsorption, which are equal in an adult whose bones have reached their final size.

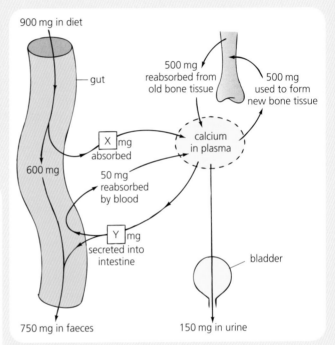

900 mg in diet

gut

500 mg reabsorbed from old bone tissue

500 mg used to form new bone tissue

X mg absorbed

calcium in plasma

600 mg

50 mg reabsorbed by blood

Y mg secreted into intestine

bladder

750 mg in faeces

150 mg in urine

Figure 15.6

a) State the figures that should have been inserted in boxes X and Y. (2)

b) Assume that a 13-year-old girl also consumes 900 mg of calcium each day. Suggest a way in which her overall calcium turnover would differ from that of an adult. (1)

9 Barium sulphate is a chemical that does not allow X-rays to pass through it. The progress of a meal containing barium sulphate can therefore be followed through the alimentary canal using X-rays to detect its location.

Table 15.2 shows the arrival and departure times of a barium meal for certain regions of the alimentary canal of six hospital patients (A–F). This information was then used to calculate the time spent by the food in each region. Some of the data are given in Table 15.3.

Patient	Time on 24-hour clock			
	Arrival of meal in stomach	Departure of meal from stomach	Departure of meal from small intestine	Arrival of meal in rectum
A	06.00	08.19	11.40	23.42
B	06.10	08.05	10.22	22.57
C	06.20	08.20	11.20	22.20
D	06.30	08.27	11.22	22.38
E	06.40	08.52	11.32	21.46
F	06.50	08.57	11.50	23.55

Table 15.2

a) What name is given to the muscular activity that moves food through the alimentary canal? (1)
b) Calculate the average time spent by a barium meal in the stomach of the patients in this survey. (1)

Patient	Time (in minutes) spent by meal:		
	In stomach	In small intestine	In large intestine
A	139	[V]	722
B	115	137	[W]
C	120	180	660
D	117	175	676
E	132	[X]	[Y]
F	127	173	725
Average		171	[Z]

Table 15.3

c) i) Convert patient C's data in the second table from minutes to hours.
   ii) Present this information for the three alimentary canal regions as a pie chart. (4)
d) Calculate the times in minutes that have been omitted from boxes V, W, X, Y and Z in Table 15.3. (5)

16 Effects of lifestyle choices

1 Table 16.1 shows the results of a survey involving thousands of British 15-year-olds. It compares certain aspects of the lifestyle of 15-year-olds in 1962 with those in 2012.

Aspect of lifestyle	1962	2012
Main items in diet	Potatoes, bread, cereals, milk, meat, eggs, fresh fruit and vegetables	Ready meals, processed food, burgers, chicken, crisps, chocolate and soft drinks
Main source of vitamin C	Fresh fruit	Sugary soft drinks
Typical leisure activities	Outdoor games, sports and limited television viewing	Computer games, internet and much television viewing
Transport to school	Almost always walked	Only about 50% walked
Body weight	Very few teenagers were overweight and obesity was rare	About 30% were overweight and obesity was common

Table 16.1

a) Among which group was obesity found to be common? (1)
b) The experts say that teenagers in 2012 were not necessarily eating more than those in 1962. With reference to leisure activities, suggest why teenagers in 2012 were often gaining weight. (2)
c) With reference to main items in the diet, suggest why teenagers in 2012 were more likely to gain weight than those in 1962. (1)
d) Suggest why it is better to get your supply of vitamin C from a fresh orange rather than from a can of sugary soft drink. (1)

e) In what THREE ways could an obese 15-year-old alter their lifestyle in order to become lighter, fitter and healthier? (3)
f) Why were *thousands* of 15-year-olds included in the survey instead of just a few? (1)

2 Stress can affect a person's health in many ways. Some of these are given in the following list: *anxiety, constipation, depression, diarrhoea, headache, inability to cope, inability to show feelings, indigestion, itchy skin, short temper.*
a) Group these effects under the headings 'physical effects' and 'mental effects'. (2)
b) What advice would you give to a friend whose job is so stressful that they are suffering from many of these effects? (1)

3 The data in Table 16.2 show the maximum values for recommended salt intake at different ages. The data in Table 16.3 refer to the food consumed at breakfast and lunch by three teenagers X, Y and Z.

Age (years)	Daily recommended maximum amount of salt (g)
1–3	2
4–6	3
7–10	5
11 and over	6

Table 16.2

a) i) What is the recommended maximum intake of salt per day for teenagers?
ii) Calculate the total mass of salt taken in by each of persons X, Y and Z so far.
iii) Who has already exceeded the recommended daily limit? (5)
b) How many different foods in the table would be described as having:
i) a high salt content?
ii) a low salt content? (2)

Person	Type of food consumed	Salt content of food (g/100g)	Mass of food consumed (g)
X	Muesli	0.4	50
	Milk	0.1	100
	Bread (toasted)	1.6	50
	Butter (salted)	2.0	20
	Sausages	2.8	100
	Beans	0.7	100
	Crisps	1.6	50
Y	Cornflakes	3.0	50
	Milk	0.1	100
	Bread	1.6	25
	Butter (salted)	2.0	10
	Bacon	5.0	50
	Chicken nuggets	1.6	100
	Pear	trace	150
Z	Milk	0.1	100
	Rice crispies	2.6	50
	Bread	1.6	25
	Butter (unsalted)	trace	10
	Poached egg	0.2	50
	Cheeseburger	1.9	200
	Apple	trace	200

Table 16.3

4 The steps in the procedure used to measure length of recovery time after vigorous exercise are given below but in the wrong order.
  A Exercise vigorously for 3 minutes.
  B Calculate the time required for pulse rate to return to normal.
  C Measure normal pulse rate before starting to exercise.
  D Take pulse rate at 1-minute intervals until normal pulse is recorded.
  a) Arrange the four steps into the correct order. (1)
  b) One afternoon this procedure was used to compare the recovery times of several college students. Some students had been playing sports at lunchtime and had missed lunch. Some students had been studying until late the night before while others had gone to bed early.
    i) Explain why the investigation is not a valid test.
    ii) Describe how it could be adapted to make it valid. (3)

5 Figure 16.1 shows a risk chart for heart disease where each black dot counts as one point of risk factor. To work out a person's total score, you add together their score from each column. In each of the questions that follow, choose the one correct answer.
  a) i) A 41-year-old man who eats a diet containing much butter and fried food is 9 kg overweight. He also smokes 30 cigarettes every day. His total score in risk points is:
    A 15
    B 16
    C 18
    D 19
    ii) The risk of the same 41-year-old man suffering heart disease is:
    A average
    B moderate
    C high
    D very high. (2)
  b) i) A 16-year-old girl who does not smoke is 2 kg over her correct weight. She eats a diet low in fat and only has fried food very rarely. Her total score in risk points is:
    A 0
    B 1
    C 2
    D 3
    ii) The risk of the same 16-year-old girl suffering heart disease is:
    A very low
    B well below average
    C below average
    D average. (2)
  c) A man who smoked 35 cigarettes daily, ate a diet containing a small quantity of butter and fried food and was 5 kg overweight was found to have a total risk score of 15 points. His age (in years) must have been in the range
    A 21–30
    B 31–40
    C 41–50
    D 51–60 (1)

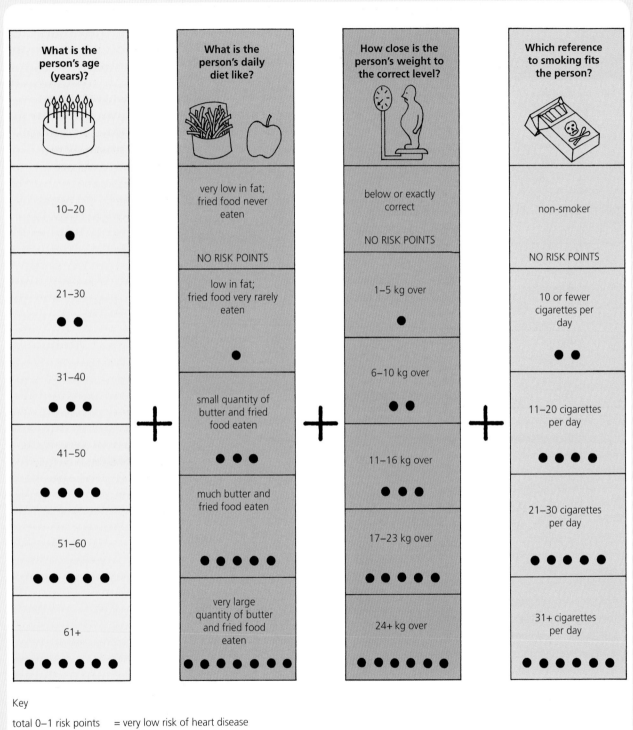

What is the person's age (years)?

10–20	●
21–30	● ●
31–40	● ● ●
41–50	● ● ● ●
51–60	● ● ● ● ●
61+	● ● ● ● ● ●

What is the person's daily diet like?

very low in fat; fried food never eaten — NO RISK POINTS

low in fat; fried food very rarely eaten — ●

small quantity of butter and fried food eaten — ● ● ●

much butter and fried food eaten — ● ● ● ●

very large quantity of butter and fried food eaten — ● ● ● ● ● ●

How close is the person's weight to the correct level?

below or exactly correct — NO RISK POINTS

1–5 kg over — ●

6–10 kg over — ● ●

11–16 kg over — ● ● ●

17–23 kg over — ● ● ● ● ●

24+ kg over — ● ● ● ● ● ●

Which reference to smoking fits the person?

non-smoker — NO RISK POINTS

10 or fewer cigarettes per day — ● ●

11–20 cigarettes per day — ● ● ● ●

21–30 cigarettes per day — ● ● ● ● ●

31+ cigarettes per day — ● ● ● ● ● ●

Key

total 0–1 risk points	= very low risk of heart disease
2–4 risk points	= well below average risk
5–7 risk points	= below average risk
8–10 risk points	= average risk
11–13 risk points	= moderate risk
14–17 risk points	= high risk
18+ risk points	= very high risk

Figure 16.1

6 Table 16.4 shows a set of results from an experiment where two girls had their pulse rates taken before, during and after a period of exercise.

Time from start (min)	Condition	Pulse rate (beats/min)	
		Jane	Karen
0	Body at rest	76	68
5	Exercise starting now	76	68
10	Exercise continuing	124	106
15	Exercise stopped now	124	106
20	Body at rest	108	68
25	Body at rest	92	68
30	Body at rest	76	68
35	Body at rest	76	68

Table 16.4

a) Plot the data as two line graphs on the same sheet of graph paper. (4)
b) What was Jane's resting pulse rate? (1)
c) What was Jane's pulse rate after 10 minutes of exercise? (1)
d) How long did it take for Jane's pulse rate to return to normal after exercise had stopped? (1)
e) What was Karen's resting pulse rate? (1)

f) What was Karen's pulse rate after 10 minutes of exercise? (1)
g) How long did it take for Karen's pulse rate to return to normal after exercise had stopped? (1)
h) i) Who recovered more quickly after exercise?
   ii) Which girl is less fit? (2)

7 In the 1990s, a survey was carried out in several countries to find out how many people per 100 000 of the population died from heart disease each year. In Japan, 10 women and 52 men died of heart disease, whereas in Germany the numbers were higher at 64 women and 246 men per 100 000. Scotland had 140 women and 510 men dying of heart disease while France had 35 female and 115 male deaths per 100 000 population.

a) Present the information in the passage as a table. (3)
b) How many more men than women died of heart disease per 100 000 population in France? (1)
c) How many more Scottish men than French men died of heart disease per 100 000 population? (1)
d) i) What was the total number of deaths per 100 000 population for Japan?
   ii) What was the total number of deaths per 100 000 population for Germany?
   iii) By how many times was Germany's total number of deaths per 100 000 population greater than that of Japan? (3)
e) What is the ratio of Scottish women to Japanese women that died of heart disease? (1)

8 The graph in Figure 16.2 charts the average number of new cases of malignant skin cancer diagnosed annually in the UK in recent times.

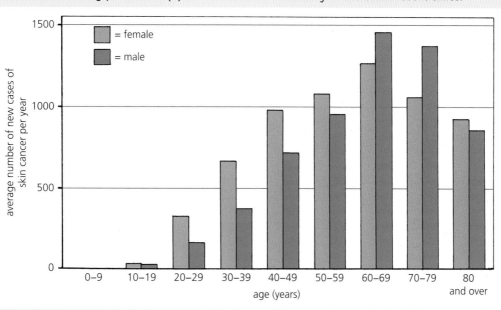

Figure 16.2

a) Which sex shows a higher incidence of skin cancer among people:

    i) under the age of 59?

    ii) between the ages of 60 and 79? (2)

b) What is the overall trend in the annual number of new cases among people aged 10 to 69? (1)

c) **i)** What is the overall trend in the annual number of cases among men aged 60 to 80 and over?

    ii) The RATE of incidence for males aged 60 to 64 is 42 per 100 000 population and for males aged 80 to 84 is 78 per 100 000 population. Explain why this information does not contradict your answer to part **i)**. (3)

9 Figure 16.3 shows five conversation bubbles containing opinions expressed by people debating the possible financing of healthcare in the UK.

    a) Which bubble is the odd one out? (1)

    b) Justify your choice. (1)

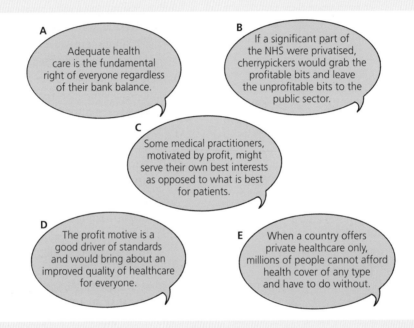

Figure 16.3

Unit 3

Life on Earth

17 Biodiversity and the distribution of life

1 The graph in Figure 17.1 shows the range of pH within which each of six soil animals is found to occur.

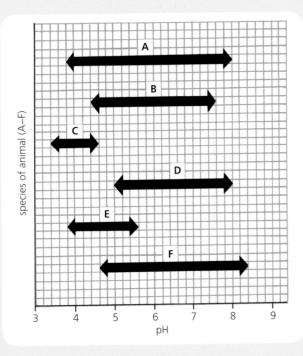

Figure 17.1

a) Which species occurs over the widest range of pH conditions? (1)
b) Which species appears to be least tolerant of acidic conditions? (1)
c) Which species is most tolerant of alkaline conditions? (1)
d) What is the lowest pH at which species F can survive? (1)
e) What is the highest pH at which species B is found? (1)
f) Which species can survive in conditions of pH below 3.75? (1)
g) How many species can tolerate pH 4? (1)
h) Which of the following pH values can be tolerated by the largest number of species?
  A 3.5
  B 4.5
  C 5.5
  D 6.5 (1)

2 Table 17.1 shows the number of different species of flowering plant and snail present in three areas of North America.

Area	Latitude		Flowering plants	Snails
1	50°	Decreasing distance from equator ↓	650	30
2	40°		1625	91
3	30°		2100	172

Table 17.1

a) What effect does latitude have on the number of different species of flowering plant and snail present? (1)
b) Suggest a reason for the trend you gave as your answer to a). (2)

3 The kite diagram in Figure 17.2 charts the results from a survey of the range and abundance of seven plants and animals present on a rocky seashore. Choose the one correct answer for each of the questions that follow.
a) Which line in Table 17.2 is correct? (1)

	Location of greatest abundance	
	Toothed wrack	**Bladder wrack**
A	Low-tide mark	Mid-tide mark
B	Mid-tide mark	Low-tide mark
C	Mid-tide mark	Mid-tide mark
D	Low-tide mark	Low-tide mark

Table 17.2

b) Which organism appears to be LEAST able to tolerate long periods out of water? (1)
  A kelp
  B lichen
  C toothed wrack
  D star barnacle

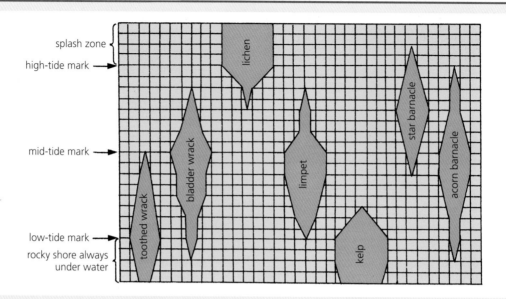

Figure 17.2

c) Which organism is able to survive the widest range of environmental conditions on this rocky shore? (1)

   A   kelp

   B   limpet

   C   bladder wrack

   D   acorn barnacle

4 The map in Figure 17.3 shows a region of coastline close to where a giant oil tanker was wrecked at sea. Prior to the disaster, the shallow waters of the coastline provided a rich source of edible crabs. Oil does not kill the crabs but harms their flesh, making them unsaleable. The extent of the shaded sector at each sample site represents the proportion of crabs with damaged flesh after the disaster.

a) Which sample site had the highest number of crabs? (1)

b) In which sample site were the crabs only rarely found? (1)

c) Describe the abundance level of crabs at sample site T? (1)

d) Name the agent of pollution that affected the crabs. (1)

e) i) In which sample site were most crabs affected relative to the population size?

   ii) Suggest why. (2)

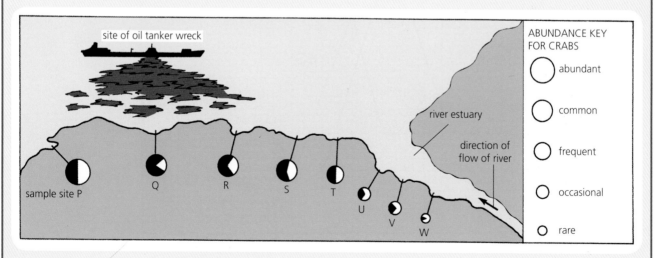

Figure 17.3

f) i) In which sample site were fewest crabs affected?
   ii) Give TWO possible reasons to explain your answer to i). (3)

5 An analysis of the pellets littering the ground near a barn owl's roost allows scientists to identify the bird's prey, as shown in Table 17.3.

Prey animal	Number present in large sample of owl pellets
Field mouse	32
Shrew	16
Vole	12
Small bird	4

Table 17.3

a) Why do owls make pellets? (1)
b) Present the information in the table as a pie chart. (3)
c) Construct a food chain which includes a barn owl. (1)
d) Some farmers shoot barn owls on sight, claiming that they are a threat to their poultry.
   i) Does the information given in the table support this view?
   ii) Explain your answer. (2)
e) Other farmers welcome the presence of barn owls on their land. Suggest why. (1)

6 The graph in Figure 17.4 shows the results from a study of the populations of two organisms, a predator and its well-fed prey, over a period of several weeks.

a) How many weeks did the population of prey organisms take to reach its maximum size? (1)
b) How many weeks did the population of predators take to go from maximum numbers to minimum? (1)
c) By how many times was the total number of prey present at week 3 greater than that present at the start of the study? (1)
d) What was the total number of predators present at week 6? (1)
e) Account for
   i) the decrease in prey
   ii) the increase in predators during the period between week 4 and week 5. (2)

7 Figure 17.5 combines a line graph of temperature with a bar graph of rainfall for a tropical rainforest. Table 17.4 gives the equivalent data for a deciduous forest.

a) Present the data for a deciduous forest in the same format as Figure 17.5 (5)
b) Compare your graph with Figure 17.5 and state THREE ways in which they differ. (3)
c) Which of the following terms can be used to refer to a major ecosystem such as a tropical rainforest? (1)
   A biome
   B niche
   C community
   D habitat

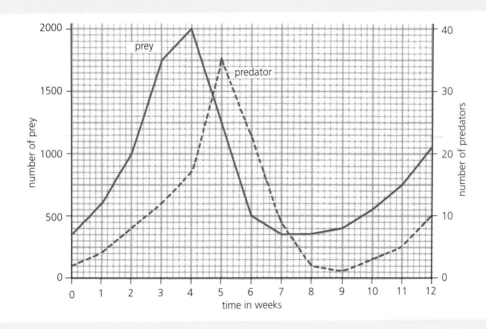

Figure 17.4

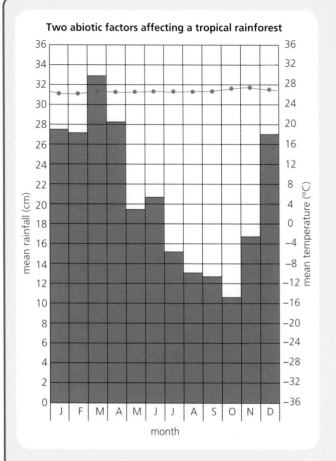

Two abiotic factors affecting a tropical rainforest

Figure 17.5

Month	Mean rainfall (cm)	Mean temperature (°C)
J	12	6
F	10	6
M	13	10
A	9	16
M	9	20
J	8	26
J	10	28
A	8	26
S	7	24
O	6	18
N	9	12
D	10	6

Table 17.4

8 Figure 17.6 shows the range of net productivity of four biomes.

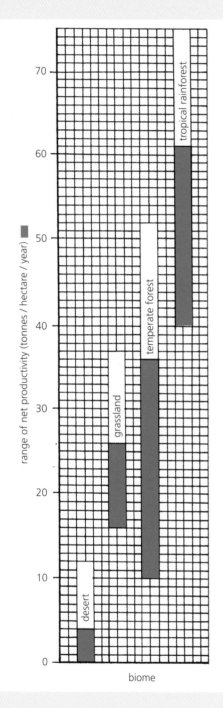

Figure 17.6

a) i) Give grassland's best annual net productivity.
   ii) Give grassland's poorest annual net productivity. (2)
b) Which biome shows the greatest range of net productivity? (1)

c) i) Which biome shows the greatest net productivity?
 ii) Which biome shows no net productivity in some years? (2)
d) Give an abiotic factor other than temperature that is largely responsible for these differences in range of net productivity. (1)

9 Copy Table 17.5 and complete the blanks using the following answers: *alder plant, brown, competition, consumer, disease, fresh, herbivorous, mole, moorland, nocturnal, predator, preyed, producer, salt, trees.* (14)

Organism	Habitat	Niche
Bracken	Open countryside	Producer that rapidly spreads by growth of underground stems; dominates environment by choking out rivals; produces poison that deters grazing animals
_____	Exposed river banks	_____ whose roots resist underwater rot and survive in water-logged soil; able to exploit environment unavailable to other _____
Red deer	Woodland and _____	_____ consumer showing population increase in absence of its extinct natural _____ (wolf)
Common seal	_____ water and sand banks	Fish-eating _____ with few serious rivals; temporary drop in numbers in recent years caused by viral _____
_____ trout	_____ water	Insect-eating consumer suffering intense _____ from introduced rainbow trout
_____	Underground burrow	_____ worm-eating consumer _____ on by owls

Table 17.5

10 Figure 17.7 shows a rocky slope at the seashore bearing star barnacles and acorn barnacles. An experiment was carried out on three similar stretches of rocky slope as follows. On the first rocky slope the barnacles were left undisturbed, the second slope was kept clear of star barnacles and the third slope was kept clear of acorn barnacles.

A survey of the distribution of the barnacles was carried out after 1 year using the abundance scale shown in Table 17.6. The results of the survey are shown in the kite diagram in Figure 17.8.

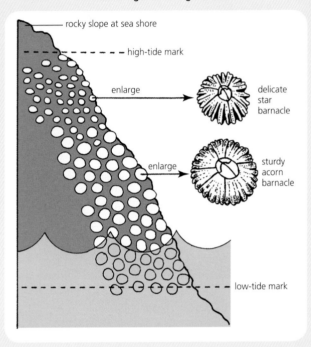

rocky slope at sea shore
high-tide mark
enlarge — delicate star barnacle
enlarge — sturdy acorn barnacle
low-tide mark

Figure 17.7

Abundance level	Abundance scale number
Rare	1
Occasional	2
Frequent	3
Common	4
Abundant	5

Table 17.6

a) i) What is the abundance scale number and level for acorn barnacles at 2 m from the low-tide mark on rocky slope 1?
 ii) What is the abundance scale number and level for star barnacles at 3 m from the low-tide mark on rocky slope 1?
 iii) Between which two distances from the low-tide mark would star barnacles and acorn barnacles be found living side by side? (3)

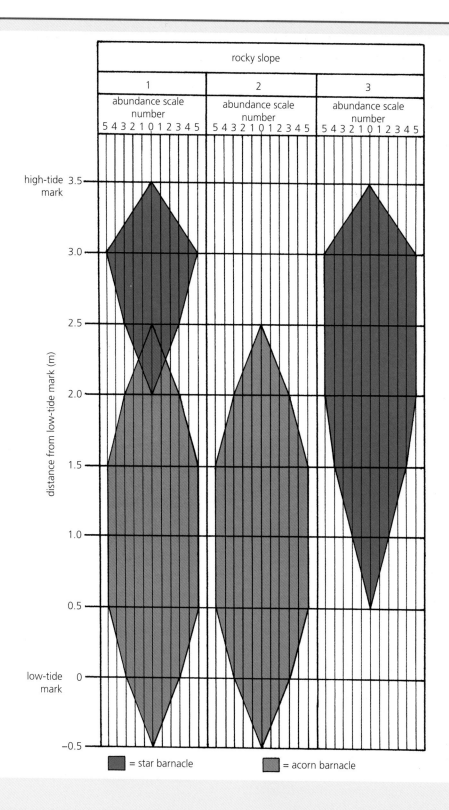

Figure 17.8

b) i) Which type of barnacle was removed from rocky slope 2?

 ii) Suggest why the remaining type of barnacle was unable to colonise the vacated territory. (2)

c) i) Which type of barnacle was removed from rocky slope 3?

 ii) Suggest why the remaining type of barnacle was able to colonise the vacated territory. (2)

18 Energy in ecosystems

1 The graph in Figure 18.1 shows the distribution of three species of organism at different depths in a loch where they form a food chain.

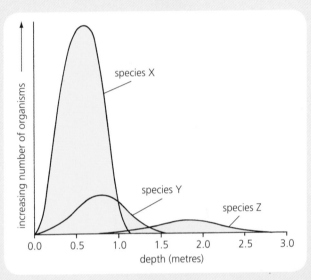

Figure 18.1

a) Which species has the widest vertical distribution in the loch? (1)

b) i) Which species is the producer?
   ii) Give TWO reasons to support your answer. (3)

c) Identify the primary and secondary consumers. Explain your choice in each case. (2)

2 a) Arrange each of the following groups of organisms into a food chain:
   i) owl, oak tree, woodmouse
   ii) salmon, plant plankton, fish louse, crustacean
   iii) animal plankton, tuna, algae, anchovy
   iv) heather, eagle, mountain hare. (4)

b) Match each of these food chains with one of the following ecosystems: *moorland, ocean, sea loch, natural woodland*. (3)

c) Match each food chain with one of the lettered pyramids in Figure 18.2. (3)

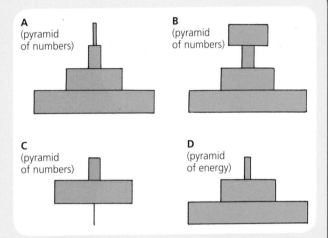

Figure 18.2

3 Read the passage and answer the questions that follow it.

Energy is lost at each link in a food chain. A certain mass of grain can support many people directly by providing them with cereal products (for example bread). On the other hand the same quantity of grain can support far fewer people indirectly by first feeding it to livestock and then providing humans with meat products.

The efficiency with which animals convert food into their own body tissues varies enormously. Carnivores are more efficient than herbivores since they consume an easily digested diet rich in protein. In addition, carnivores do not depend on the presence of energy-consuming micro-organisms in their gut to aid digestion of plant cell walls and their faeces contain less undigested material.

If an animal is endothermic ('warm-blooded'), it maintains its body temperature at around 37 °C and therefore uses much of the energy in its food to stay warm in cold weather. Ectothermic ('cold-blooded') animals do not maintain a body temperature above that of the environment and therefore more of the energy in their food is available for secondary productivity.

a) Allowing a 10% conversion rate each time, state the biomass of human tissue that could be produced from 1000 kg of cereal as:
   i) bread
   ii) pork chops from pigs fed on cereal. (2)
b) Table 18.1 refers to two animals X and Y.

Animal	Percentage energy absorbed from food	Percentage energy built into tissues
X	36.6	5.2
Y	84.3	29.8

Table 18.1

   i) Which animal is the salmon and which is the sheep? Give TWO reasons to support your choice.
   ii) What happens to the energy not converted into the animals' body tissues? (4)
c) Explain why intensive farming methods often include:
   i) keeping animals indoors, especially during winter
   ii) rearing animals in tightly confined spaces. (4)
d) i) Which of the following forms of Scottish agriculture would you expect to be more productive in terms of energy conversion into animal tissue per gram of feed: trout fish farm or deer farm?
   ii) Explain your choice of answer. (2)
4 Table 18.2 below gives the results from a plant competition experiment where five groups of pea plants were grown in areas of similar fertile soil measuring 0.25 m².

Number of plants per 0.25 m²	Average number of pods per plant	Average number of seeds per pod
20	8.3	6.0
40	6.8	5.9
60	3.9	6.2
80	2.7	5.9
100	2.1	6.0

Table 18.2

a) Plot the data in the table as two line graphs using graph paper similar to that in Figure 18.3, which shows how to lay out the axes. (2)

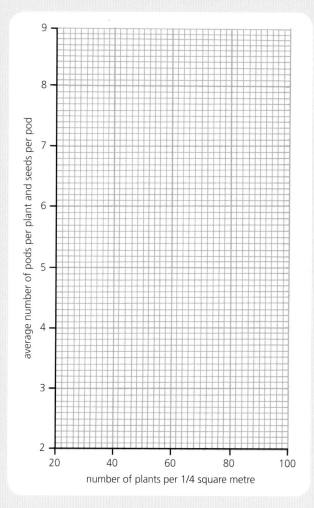

Figure 18.3

b) i)   Which feature appears to be affected by competition?
   ii) In what way is this feature affected as density of plants in 0.25 m² increases?
   iii) Suggest TWO factors that neighbouring pea plants may be competing for. (4)
c) Calculate the total number of seeds produced by
   i) 20 plants on 0.25 m²
   ii) 100 plants on 0.25 m². (2)
d) i)   Since seed mass is found to be unaffected by competition, which number of plants, 20 or 100, would be the better number to grow per 0.25 m²?
   ii) Explain your answer. (2)

e) It is possible that these results are unusual and not typical of pea plants in general. What should now be done to check the reliability of these results? (1)

5 Copy and complete the flow diagram in Figure 18.4 using the following answers. (7)
- nitrogen-fixing bacteria in root nodules
- denitrifying bacteria
- nitrifying bacteria type 1
- nitrifying bacteria type 2
- nitrogen gas in air
- nitrate in soil
- ammonium compounds
- nitrite in soil

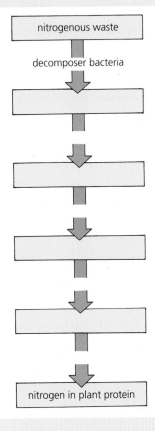

Figure 18.4

6 A student set up an experiment to investigate the effect of competition on the growth of a population of cress seedlings. Figure 18.5 shows a simplified version of his set-up after 5 days in a seed propagator. Table 18.3 gives his results.

a) State the means by which the factor under investigation was varied. (1)

b) State THREE factors that the diagram shows were kept constant. (3)

c) Copy and complete Table 18.3. (2)

d) Plot a line graph of number of seeds planted against percentage number of healthy seedlings with green leaves. (3)

e) Make a generalisation from your graph about the effect of competition on the growth of a population of cress seedlings. (1)

f) Suggest ONE factor other than water that cress seedlings in carton E could be competing for. (1)

g) Figure 18.6 is a partly completed diagram of an alternative method of investigating the effect of competition on cress seedlings. Make a simple diagram of plates B and C to show them set up and ready at the start of the experiment. (2)

Number of seeds planted	Number of healthy seedlings with green leaves	% number of healthy seedlings with green leaves
100	87	
200	178	
300	204	
400	228	
500	225	

Table 18.3

tall healthy seedling

2 g cotton wool + 20 cm³ water in each carton

sickly yellow seedling

	A	B	C	D	E
number of seeds planted	100	200	300	400	500

Figure 18.5

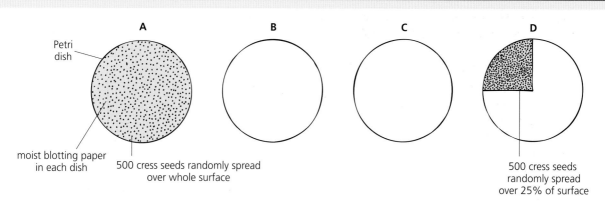

Petri dish

moist blotting paper in each dish

500 cress seeds randomly spread over whole surface

500 cress seeds randomly spread over 25% of surface

Figure 18.6

7 *Paramecium* (see Figure 18.7) is a unicellular animal that feeds on bacteria. Two closely related species are *Paramecium caudatum* and its smaller relative *Paramecium aurelia*. Graphs A and B in Figure 18.8 represent the growth of a population of *P. caudatum* and one *of P. aurelia* each cultured alone in favourable conditions. Graph C shows the growth curve that results when the two species are cultured together.

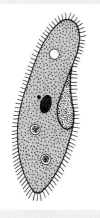

Figure 18.7

a) Suggest THREE factors that would contribute to the 'favourable conditions' referred to in the passage. (3)

b) For what were the two species of *Paramecium* probably competing in the part of the experiment represented by graph C? (1)

c) What is the maximum number of each type of *Paramecium* produced when it is cultured alone? (2)

d) On which day in graph C did *P. caudatum* go into decline as a result of competition? (1)

e) i) Was the growth of the population of *P. aurelia* affected by competition?
   ii) Explain your answer. (2)

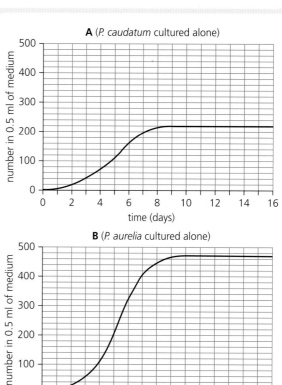

A (*P. caudatum* cultured alone)

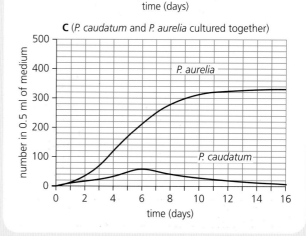

B (*P. aurelia* cultured alone)

C (*P. caudatum* and *P. aurelia* cultured together)

P. aurelia

P. caudatum

Figure 18.8

f) i) Predict the day on which *P. caudatum* would be completely wiped out if the current trend were to continue.

ii) Would this be the result of intraspecific or interspecific competition? (2)

8 The graph in Figure 18.9 shows the economics of defending a territory of varying size by a certain species of bird.

a) State the range of territory size that can be realistically defended by this species of bird. (1)

b) Suggest why the bird cannot survive in a territory of 10 m². (1)

c) Give ONE example of the high 'cost' that makes the defence of a territory of 140 m² unrealistic for this type of bird. (1)

d) i) What is the optimum size of territory for this species of bird?

ii) Explain why. (2)

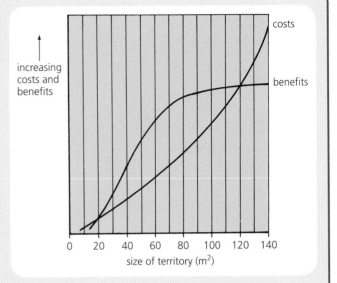

Figure 18.9

19 Sampling techniques and measurements

1 The lawn in Figure 19.1 measures 10 metres in length by 10 metres in breadth.

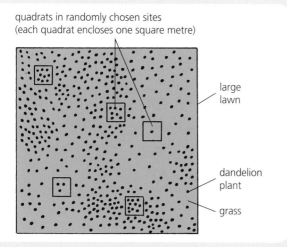

quadrats in randomly chosen sites
(each quadrat encloses one square metre)

large lawn

dandelion plant

grass

Figure 19.1

a) Calculate the area of the lawn. (1)
b) Estimate the total number of dandelion plants growing on the lawn using only the information provided by the randomly chosen quadrats. (1)
c) The owner decided to try to remove the dandelions by spraying the lawn with selective weedkiller. Figure 19.2 shows the lawn several weeks after spraying. Using the same method, estimate the total number of dandelion plants that are now growing on the lawn. (1)

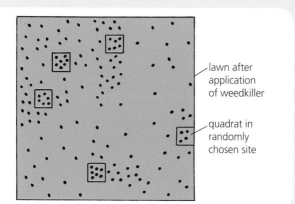

lawn after application of weedkiller

quadrat in randomly chosen site

Figure 19.2

d) i) From a consideration of your results so far, does the weedkiller appear to have been effective?
   ii) Explain your answer. (2)
e) Taking an overall view of the lawn before and after spraying, the weedkiller does appear to have been at least partly successful.
   i) State the source of error in the sampling technique that was responsible for failing to show up this difference.
   ii) Explain how this error could be reduced to a minimum in a future investigation using the same sampling technique. (2)

2 The apparatus shown in Figure 19.3 is used to extract microscopic organisms that live in the film of water around soil particles.

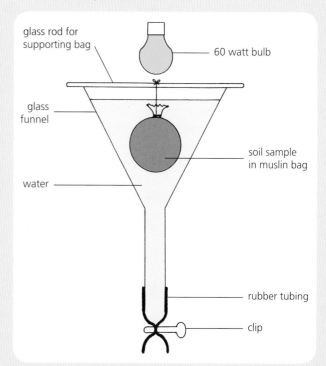

glass rod for supporting bag

60 watt bulb

glass funnel

soil sample in muslin bag

water

rubber tubing

clip

Figure 19.3

a) Explain how it works. (2)
b) By what means could samples of the aquatic organisms from this soil sample be prepared for viewing under a microscope? (1)
c) Why is a Tullgren funnel not suitable for extracting aquatic soil organisms? (1)

3 Using a variety of equipment and sampling techniques, scientists investigated the state of five environmental factors at different depths in a fresh-water loch. Their results are presented by the kite diagram in Figure 19.4. In the two questions that follow, choose the one correct answer.

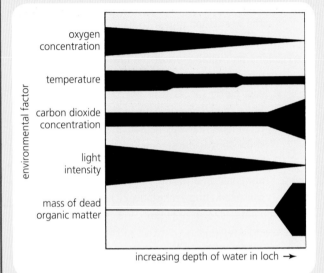

increasing depth of water in loch →

Figure 19.4

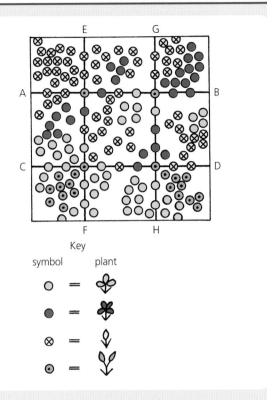

Key

symbol   plant

Figure 19.5

Figure 19.6

a) Algae are found growing close to the surface of the loch. Which of the following is the factor that restricts the distribution of the algae to this region of the water? (1)

  A concentration of oxygen
  B concentration of carbon dioxide
  C intensity of light
  D mass of dead organic matter

b) Most of the loch's population of bacteria are found at the bottom. Which of the following is the factor mainly responsible for this distribution of bacteria? (1)

  A concentration of oxygen
  B temperature
  C concentration of carbon dioxide
  D mass of dead organic material

4 Figure 19.5 shows a piece of ground viewed from above. It was sampled by taking four line transects (AB, CD, EF and GH) and recording the type of plant present at regular intervals along each transect.

a) Which of the four transects corresponds to Figure 19.6? (1)

b) From the results of this one transect alone, which plant type seems to be most abundant? Give its colour. (1)

c) Look again at the diagram of the piece of ground. Is the plant that you gave as your answer to b) in fact the most abundant? (1)

d) Identify a more accurate method of estimating which plant type is the most abundant in the area. (1)

5 During an investigation of an ecosystem, pupils carried out tree-beating and caught the six animals shown in Figure 19.7. Figure 19.8 shows an incomplete version of the branched key that they constructed.

a) Copy and complete the branched key. (4)

b) Use the key to identify animals A, B, C, D, E and F. (1)

c) Convert the branched key into a key of paired statements. (3)

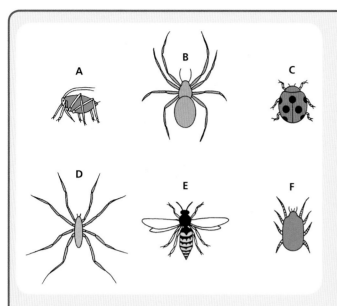

A

B

C

D

E

F

Figure 19.7

6 Two boys were asked to investigate the effect of light intensity on the distribution of meadow buttercup plants at the edge of the oak wood shown in Figure 19.9. They pegged out a string line from point X outside the wood to point Y inside the wood. Along this line transect at points 1–10 they placed a metre-square quadrat and counted the number of meadow buttercup plants present in each quadrat. This number was given an abundance score as shown in Figure 19.10.

The boys also measured the light intensity falling on each quadrat by taking one reading using a light meter, which gave readings on an 8-point scale of A–H where A = dimmest and H = brightest. Their results are shown in Figure 19.11.

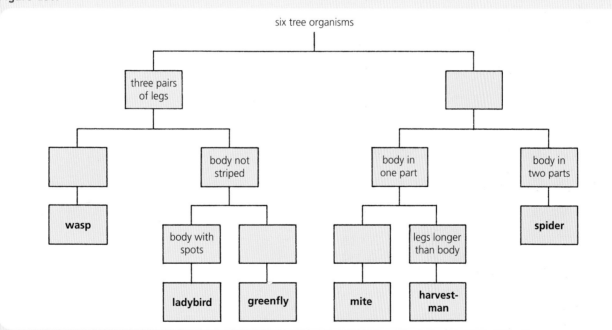

six tree organisms

three pairs of legs

body not striped

body in one part

body in two parts

wasp

body with spots

ladybird

greenfly

legs longer than body

mite

harvest-man

spider

Figure 19.8

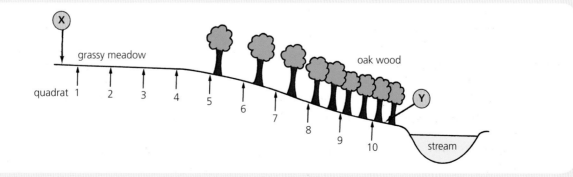

X

grassy meadow

oak wood

quadrat 1 2 3 4 5 6 7 8 9 10

stream

Figure 19.9

number of buttercup plants	abundance score	symbol
26 or more	abundant	
11–25	frequent	
6–10	occasional	
1–5	rare	
0	absent	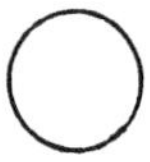

Figure 19.10

quadrat	1	2	3	4	5	6	7	8	9	10
light intensity	H	H	H	H	F	E	D	C	C	C
abundance of meadow buttercups									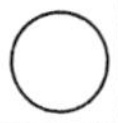	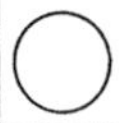

Figure 19.11

Time on 24-hour clock	Temperature of soil at 3 cm depth (°C)	Temperature of soil at 30 cm depth (°C)
09.00	7.5	8.1
12.00	8.0	8.0
15.00	8.7	7.9
18.00	8.7	8.0
21.00	8.6	8.0
24.00 (00.00)	8.3	8.1
03.00	8.0	8.0
06.00	7.5	7.9
09.00	7.5	8.0

Table 19.1

a) What abiotic factor was measured in the investigation? (1)

b) In what way does the abiotic factor gradually change along the line transect from quadrat 1 to 10? (1)

c) i) In what way does the abundance of buttercup plants change along the line transect from quadrat 1 to 10?

  ii) What relationship appears therefore to exist along the transect between abundance of buttercups and light intensity?

  iii) Suggest a possible reason for this apparent relationship. (3)

d) i) Considering that some parts of the ground at the edge of a wood on a sunny day are lit up by patches of bright sunlight whereas others are in the shade, spot a shortcoming in technique used by the boys to measure light intensity.

  ii) Suggest how the shortcoming that you identified could be overcome. (2)

e) It is possible that some other abiotic factor is wholly or partly responsible for the distribution of the buttercup plants. Suggest TWO such factors. (2)

7 Table 19.1 shows the results of recording soil temperatures at two different depths in the same soil (which lacked vegetation) during a 24-hour period in summer.

a) Present the data as two lines graphs sharing the same axes. (4)

b) How frequently were the readings of soil temperature taken? (1)

c) Suggest a suitable piece of equipment that could be used to take the readings. (1)

d) At which TWO times of day was the temperature equal at both depths of soil? (1)

e) Account for the trend in soil temperature that occurs between 09.00 and 15.00 hours for the 3 cm depth. (1)

f) Account for the trend in soil temperature that occurs between 21.00 and 06.00 hours for the 3 cm depth. (1)

g) i) Which soil depth showed less variation in temperature over the 24-hour period?

ii) Account for this lack of variation. (2)

h) i) Suggest the effect that a dense covering of plants would have on the temperature of the soil at a depth of 3 cm.

ii) Explain your answer. (2)

8 Figure 19.12 charts the results from an experiment to investigate the effect of soil pH on the distribution of four plant species (W, X, Y and Z). The soil pH and percentage cover of each plant species were recorded at each of ten sample points at regular intervals along a line transect. The plant cover data are presented as kite diagrams. In the three questions that follow, choose the one correct answer.

a) Only two of the plant species were recorded at which sample point? (1)

A 3

B 5

C 6

D 7

b) Species W, X and Y were all found to be growing in soil of which pH? (1)

A 5.1

B 5.7

C 6.1

D 6.9

c) Which species is able to tolerate the widest range of soil pH? (1)

A W

B X

C Y

D Z

9 The graph in Figure 19.13 shows the effect of speed of water movement on the numbers of two species of insect larvae, A and B, found at the surface of a river.

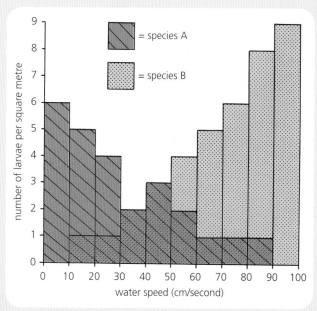

Figure 19.13

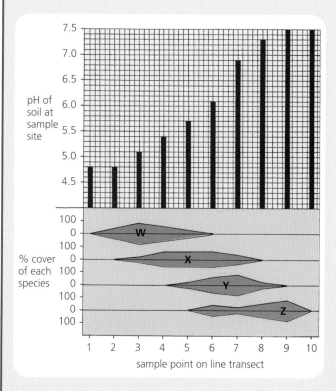

Figure 19.12

a) At which range of water speed was the greatest number of species A recorded? (1)

b) From which range of water speed was species A absent? (1)

c) At which ranges of water speed were equal numbers of species A and B found? (1)

d) For each species, state the relationship that exists between numbers present and the given ranges of water speed. (2)

e) Table 19.2 shows the sizes of inanimate objects moved by different water speeds.

i) Which species of insect larva would you expect to be more numerous in a stretch of river where the water speed is able to move particles of coarse gravel?

ii) Explain your answer. (2)

Water speed (cm/s)	Diameter of object (mm)	Example of object
10	0.2	Mud particle
25	1.3	Sand particle
50	5.0	Gravel particle
75	11.0	Coarse gravel
100	20.0	Pebble

Table 19.2

20 Adaptation, natural selection and evolution

1 Table 20.1 shows the results of an investigation into the effect of increasing doses of radiation on germinating barley grains which were planted in batches of 100.

Units of radiation (roentgens)	0	1000	2000	3000	4000	5000
Number of barley grains germinating and producing a shoot	96	32	16	10	6	2
Number of shoots with abnormal leaves	0	16	12	9	6	2
Percentage number of shoots with abnormal leaves						

Table 20.1

a) Construct a line graph to show the effect of increasing radiation on number of barley grains germinating. (3)

b) Draw a conclusion from your graph. (1)

c) Explain why batches containing as many as 100 grains were planted. (1)

d) Calculate the percentage number of shoots with abnormal leaves for each dose of radiation (as indicated by the blank boxes in the table). (2)

e) Add a second vertical scale to the right-hand side of your graph and plot the effect of increasing radiation on the percentage number of shoots with abnormal leaves as a second line graph. (2)

f) Draw a conclusion from your second line graph. (1)

g) Predict the dosage of radiation that would cause 80% of the barley grains to fail to germinate. (1)

2 Table 20.2 shows recent data giving the number of new cases of malignant skin cancer diagnosed per 100 000 population per year for all ages.

Gender	Number of new cases per 100 000 population per year				
	Scotland	England	N. Ireland	Wales	UK
Female	18.4	17.3	16.3	16.8	[X]
Male	17.3	[Y]	12.2	21.9	17.1

Table 20.2

a) Calculate the values missing from boxes X and Y. (2)

b) i) In general which gender shows a slightly higher incidence of new cases of malignant skin cancer?

   ii) Which country in the UK is the exception to this trend according to the data in the table? (2)

c) If Scotland has a population of 6 million, calculate the number of new cases of malignant skin cancer diagnosed per year among:
   i) men
   ii) women. (2)

3 Read the passage and answer the questions that follow it.

Camels lose water steadily through their skin and their breath, as well as in urine and faeces. Although camels can go for days without drinking water, they normally survive during this time by feeding on plants. The special compartments attached to a camel's stomach are filled with foul-smelling liquid (containing fermenting food) which is equal in water concentration to mammalian blood.

If a human being loses more than 12% of their body water, they are found to be in a critical state because their blood has become thick and sticky and their heart is having difficulty pumping it. A camel can lose up to 25% of its body water without straining its heart because its blood does not become thick and sticky.

A camel's hump consists of fat and acts as an energy reserve. Fat does yield metabolic water during tissue respiration but so much extra oxygen is needed to burn fat that any gains in water are lost as water vapour during the increased rate of breathing that occurs. The camel's coat provides insulation against the heat of the desert day. The animal's body temperature begins at 34 °C at dawn and does not reach 40 °C until around midday. Thus the rate of sweating is kept to a minimum during the first part of the day. The body temperature drops back down to 34 °C during the night.

a) Compared with a kangaroo rat, which behavioural adaptation is a camel unable to employ to conserve water? (1)

b) Compared with a kangaroo rat, which additional method of water loss occurs from a camel's body? (1)

c) i) Is hypothesis 1 in Figure 20.1 supported by the information in the passage? Justify your answer.

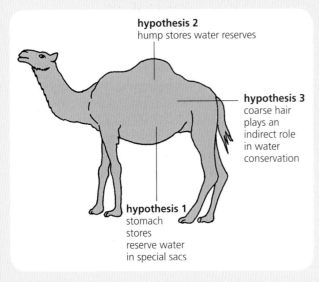

hypothesis 2
hump stores water reserves

hypothesis 3
coarse hair plays an indirect role in water conservation

hypothesis 1
stomach stores reserve water in special sacs

Figure 20.1

ii) Would a human being, almost dead from dehydration, be able to save their own life by drinking the liquid from a camel's stomach compartments? Explain your answer. (4)

d) i) Is hypothesis 2 supported by the passage? Explain your answer.

ii) Why does a camel have a hump? (3)

e) Is hypothesis 3 supported by the passage? Explain your answer. (2)

f) Explain why a camel which has lost 20% of its body water by dehydration is still capable of the same physical exertion as a well-watered camel. (1)

4 On average, a pair of foxes produce a litter of five cubs once a year. Since the offspring mature within a year and set up territories of their own, millions of foxes could be produced in a few years. Using the terms *over-production*, *competition*, *natural selection* and *variation* in your answer, explain why the world is not over-populated by foxes. (4)

5 Read the passage and answer the questions that follow it.

Two forms of the peppered moth (*Biston betularia*) exist. One form is light brown with dark speckles; the other is completely dark (melanic) in colour. They differ by only one allele of the gene controlling the formation of dark pigment (melanin). Both forms of the moth fly by night and rest on the bark of trees during the day.

In non-polluted areas, the tree trunks are covered with pale-coloured lichens and the light-coloured moth is well camouflaged against this pale background (see Figure 20.2). However, the dark form is easily seen and eaten by predators such as thrushes.

In polluted areas, toxic gases kill the lichens and soot particles darken the tree trunks. As a result the light-coloured moth is easily seen whereas the dark one is well hidden and is favoured by natural selection. However, polluted areas have undergone cleaning up campaigns in response to the Clean Air Acts (1956, 1963 and 1968). Therefore the pale form is being naturally selected at the expense of the melanic form, which is losing its selective advantage.

The graph in Figure 20.3 shows the results from a survey of lichens carried out in the vicinity of a large industrial city in Britain in the early 1950s.

a) State the relationship that exists between the distribution of lichens and melanic moths. (1)

b) Explain in evolutionary terms why the melanic moth graph takes the form shown in Figure 20.3. (2)

c) Predict the form that a graph of the pale-coloured peppered moth numbers would have taken if drawn using the same axes. (1)

d) In the centre of the same city, the percentage number of melanic moths decreased from 98% in 1955 to 75% in 1980. Account for this change. (2)

e) Was the change in the moth's environment brought about by a biotic or an abiotic factor? (1)

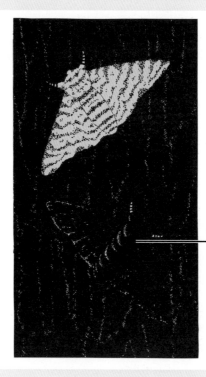

light form enjoys selective advantage on lichen-covered trunk in non-polluted area

dark form enjoys selective advantage on soot-covered trunk in polluted area

Figure 20.2

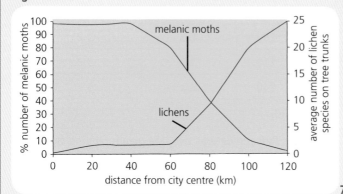

Figure 20.3

6   The data in Table 20.3 refer to Scottish cases of a type of chest infection caused by a species of bacterium. Before 1980, the bacterium was sensitive to the antibiotic erythromycin.

Year	Percentage of reported cases successfully treated by erythromycin
1984	85
1986	74
1988	62
1990	48

Table 20.3

a) Account for the trend shown in the table. (1)
b) Predict what would probably have happened by the year 1998 if doctors had continued to prescribe erythromycin. (1)
c) By the year 1998, the percentage of reported cases successfully treated had actually risen to 96%.
    i)   Suggest how doctors brought about this reversal in the trend.
    ii) Suggest why they were unable to achieve a 100% success rate. (2)

7   In the USA in the 1950s, some brands of toothpaste and chewing gum contained antibiotics. This practice has been discontinued.
a) What is the advantage in the short term of adding antibiotics to toothpaste? (1)
b) What is the disadvantage in the long term of continuing with this practice? (1)

8   St Kilda and several Hebridean islands such as Lewis, Mingulay and Rhum each possess their own distinct subspecies of the field mouse (scientific name *Apodemus sylvaticus*).
a) Suggest how each of these subspecies originated. (2)
b) i)   Under what circumstances may each evolve into a separate species?
    ii) What circumstances could prevent the process of speciation from taking place? (2)

21 Human impact on environment

1 Table 21.1 refers to intensive farming practices. Copy and complete the table using the following answers:
   A Use of pesticide spray
   B Reduction in biodiversity and poisoning of helpful insects
   C To reduce the amount of energy lost by animals and make more available for growth
   D Use of herbicide spray
   E Farm animals kept indoors
   F To prevent weeds using essential resources such as minerals intended for the crop (5)

Practice	Details of practice	Reason for practice	Possible adverse side effect(s)
Removal of competitors from area where crop is being grown			Reduction in biodiversity
Removal of insects and other pests that feed on the crop		To prevent energy present in food produced by the crop being transferred to unwanted consumers	
	'Battery' farming of animals in confined space		Increase in risk of disease and decrease in quality of animals' life

Table 21.1

2 Table 21.2 shows the results from an experiment to investigate the effect of nitrate and potassium on yield of maize plants. (ha = hectare = 100 acres)

Mass of nitrate added (kg/ha)	Yield of maize plants (tonnes/ha)	
	Treatment A (no potassium added)	Treatment B (135 kg/ha of potassium added)
0	2.0	2.9
30	2.9	4.4
60	3.8	5.8
90	4.4	7.0
120	5.9	8.1
150	6.1	8.8
180	4.5	9.2
210	3.6	9.2
240	2.3	9.2

Table 21.2

a) Plot the data as two line graphs sharing the same axes. (4)
b) What was the yield of maize when 90 kg/ha of nitrate was added in the absence of potassium? (1)
c) What was the yield of maize when 135 kg/ha of potassium was added in the absence of nitrate? (1)
d) i) What was the yield of maize when neither nitrate nor potassium were added?
   ii) How was any yield of the crop possible when neither mineral had been added as fertiliser? (2)
e) In treatment B, at what concentration is quantity of nitrate no longer the factor limiting the yield of maize? (1)
f) If a farmer lacked a supply of potassium, why would it be a waste of money to add more than 150 kg/ha of nitrate to the soil? (1)

3 The graph in Figure 21.1 refers to the quantity of nitrogen released from the soil and the quantity required by a crop planted in October.

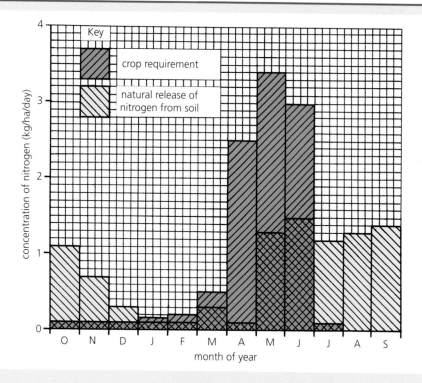

Figure 21.1

a) i) Identify the two months of the year during which the highest natural release of nitrogen occurs.

ii) Which of these peaks occurs at a time when no demand for nitrogen is being made by the crop?

iii) Which of these peaks fails to fully support the nitrogen requirement of the crop? (4)

b) i) During which ONE of the following months would application of fertiliser be most likely to lead to leaching of nitrate from this farmland into the local supply of drinking water?

    A March

    B May

    C June

    D August

ii) Explain your choice of answer to i). (2)

c) To reduce the chance of nitrate being leached out of a soil, the farmer should apply nitrogen fertiliser:

    A when the cereal crop is actively growing

    B in early autumn before the ground freezes over

    C in one large dose rather than several smaller ones

    D when the soil is bare and fallow in the winter

i) Choose ONE correct answer only.

ii) Explain why each of the others is wrong. (4)

4 The graph in Figure 21.2 shows the results of an analysis of the breast muscles of several species of water birds for a non-biodegradable pesticide.

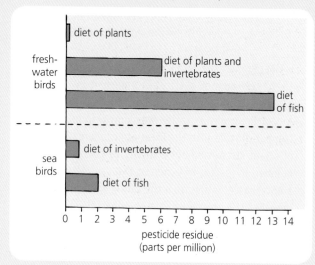

Figure 21.2

a) i) From the graph, state the general relationship that exists between a bird's diet and the concentration of pesticide residue in its muscle tissues.

ii) Account for this relationship. (3)

b) i) Which of the environments was less severely affected by the pesticide, as indicated by these data?

ii) Suggest why. (2)

5 The graph in Figure 21.3 shows the relationship between concentration of pesticide in the eggs of a predatory bird and thickness of egg shell.

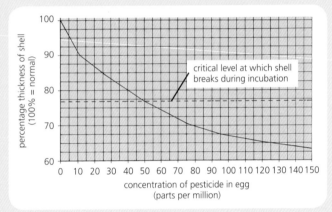

Figure 21.3

a) Describe the relationship shown by the graph. (1)

b) i) If this species of bird lays eggs containing 25 ppm of pesticide residue, by what percentage will the shell's normal thickness be reduced?

ii) How many more parts per million of pesticide would need to be present to reach the shell's critical level?

iii) Predict the bird's reproductive success when 100 ppm of pesticide are present in its eggs. Explain your answer. (4)

6 A coal-fired power station was known to release sulphur dioxide into the local atmosphere. A group of scientists carried out a survey by counting the number of species of lichen at five sample sites in a northeasterly line from the power station. They repeated the procedure in a southeasterly line, as shown in Figure 21.4. The diagram also gives details of an annual survey of wind direction for the area. Table 21.3 gives the results of the lichen survey.

Compass direction	Mean number of different species of lichen at sampling site:				
	A	B	C	D	E
NE line	2	2	3	3	4
SE line	3	5	6	8	11

Table 21.3

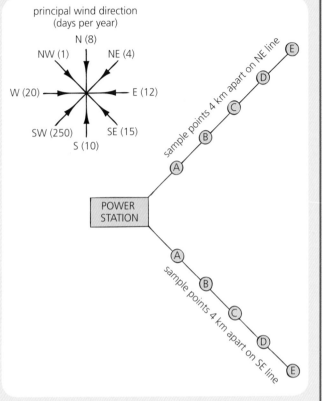

Figure 21.4

a) How far from the power station was:
   i) sample site A?
   ii) sample site E? (2)

b) i) Why were several counts taken at each sample site and a mean value calculated for the number of lichen species present?

ii) What happens to the biodiversity of lichen species as the distance from the power station increases along the southeast (SE) sample line?

iii) Explain why. (4)

c) i) How many calm days lacking significant wind occurred during the year to which the wind survey refers?

ii) In which direction did the wind blow least often?

iii) In which direction did the wind blow most often? (3)

d) i) Make a generalisation about the difference between the number of different lichen species found on the northeast (NE) line compared with the southeast (SE) line.

ii) Give a possible explanation for this difference based on the information given. (3)

7 Repeated sampling of the water in the river shown in Figure 21.5 was carried out at sample points 1–5 during the years 2001 and 2005. The results are summarised in Figure 21.6.

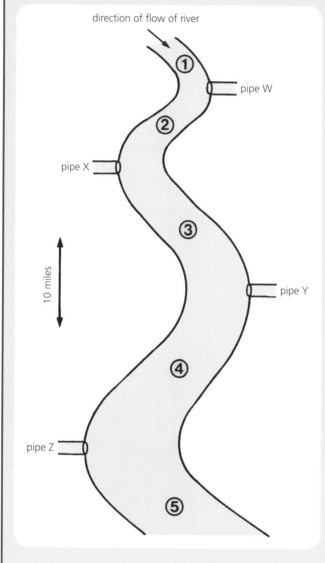

Figure 21.5

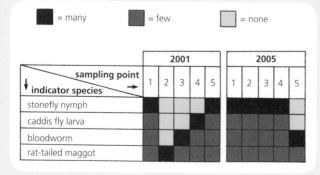

Figure 21.6

a) i) From which pipe was untreated sewage from an overloaded sewage works discharged into the river in 2001?

ii) Explain your choice of answer.

iii) Suggest a possible change that could have occurred to this sewage works during 2002–2004 to account for the result obtained in 2005. (3)

b) In 2005 a paper mill began discharging untreated organic waste into the river through pipe Z. Describe the effect this had on:

i) the type of animal most commonly found at sample site 5

ii) the oxygen concentration of the water at sample site 5. (2)

8 Read the following passage and answer the questions that follow it.

The cottony cushion scale insect first appeared in the Californian orange groves in 1868 following its accidental introduction from Australia. In the absence of natural enemies, it quickly multiplied and within 20 years was causing major damage to the citrus trees. The fruit growers were close to ruin.

Scientists studying the pest's natural enemies found that a certain type of Australian beetle (a species of ladybird) could be used to clear the orange trees of the scale insect (see Figure 21.7). Following the introduction of the predators, the pest was brought under control within a few years (see Figure 21.8) and the citrus fruit industry was saved.

Scientists monitoring populations found that the cottony cushion scale insect was continuing to survive in low numbers but was no longer posing a major threat to orange trees since it was being kept in check by ladybirds. When, at a later date, the orange groves were treated with pesticide, most of the scale insects died. However a few were resistant to the chemical and survived. All of the ladybirds died. This resulted in a new epidemic of scale insects, which was only curbed by the reintroduction of ladybirds.

a) Construct a food chain to show the relationship between the organisms mentioned in the passage. (1)

b) Study the graph in Figure 21.8 and identify which line is:

i) the predator

ii) the prey.

iii) Explain your choice in each case. (3)

c) How many years did it take for the pest population to reach a level where it was inflicting major economic damage? (1)

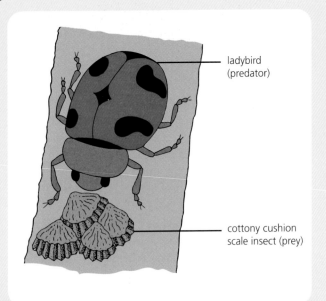

ladybird
(predator)

cottony cushion
scale insect (prey)

Figure 21.7

d) During which year were the predators first
   introduced? (1)
e) Why did the number of predators drop after
   1895? (1)
f) Describe the relationship between predator and
   prey between the years 1900 to 1940. (2)
g) Initially fruit growers were pleased with the
   outcome of the pesticide treatment carried out in
   the 1940s.
   i)   Suggest why.
   ii) Why was their delight short-lived?
   iii) By what means was the problem overcome? (3)
h) Predict the form that the lines in the graph took
   after 1960. (1)

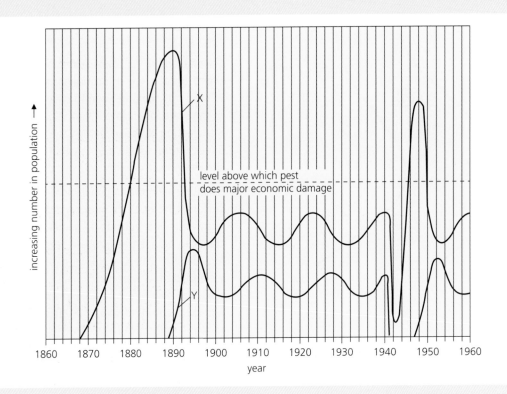

Figure 21.8

Answers

1 Cell structure

1 a) W = rhubarb epidermal cell, X = *Elodea* leaf cell,
        Y = onion epidermal cell, Z = yeast cell (3)
   b) 1 = large central vacuole, 2 = cell wall,
        3 = chloroplast, 4 = cytoplasm, 5 = cell wall,
        6 = nucleus, 7 = cell membrane, 8 = cytoplasm,
        9 = cell wall, 10 = nucleus (10)
2 See Figure An 1.1 (3)

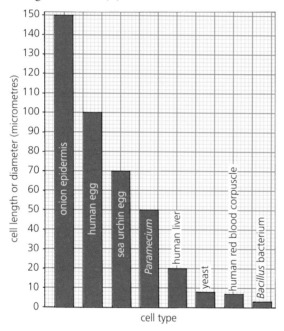

Figure An 1.1

3 A = 3, B = 1, C = 2 (2)
4 a) i) Nucleus
        ii) *Elodea* leaf cell and onion epidermal cell (2)
   b) i) Potato tuber cell
        ii) They contain structures that turn blue-black
            when iodine solution is added to them. (2)
5 a) Movement (1)
   b) It uses rotting food as a source of food. (1)
   c) It enables the organism to use light for
        photosynthesis. (1)
   d) Chloroplast (1)
   e) Cell wall (1)
   f) It has some features typical of plant cells and
        some typical of animal cells and therefore fits
        neither group exactly. (2)

6 See Figure An 1.2 (5)

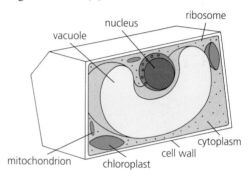

Figure An 1.2

7 a) D, F, B, E, A, C (1)
   b) i) F
        ii) D
        iii) Polish the eyepiece lens with a lens tissue. (3)
   c) C (1)

2 Transport across cell membranes

1 a) Iodine solution has reacted with starch <u>inside</u> the
        Visking tubing 'sausage'. (1)
   b) Iodine solution (1)
   c) Starch 'solution' (1)
   d) i) Set up the experiment with the two liquids
            reversed.
        ii) If there is a blue-black colour on the outside of
            the Visking tubing only, this supports the theory
            that only molecules of iodine solution are small
            enough to pass through the membrane (and
            react with the starch). If there is a blue-black
            colour on the inside of the Visking tubing only,
            this supports the theory that the membrane
            allows particles of any size to enter but not leave
            (since the starch must have entered and reacted
            with the iodine solution). (3)
2 a) i) Solution 1
        ii) Solution 5 (2)
   b) 1 = Y; 2 = X; 3 = Y; 4 = Z; 5 = X; 6 = Y (6)

3 a) See table (2)

Change in length (mm)	Percentage change in length
+8	+16
+3	+6
−2	−4
−7	−14
−12	−24

b) See Figure An 2.1 (4)

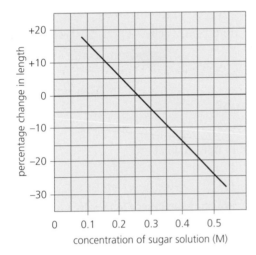

Figure An 2.1

c) i) Higher
   ii) The potato cylinder gained length, therefore
       water must have passed in from a region of
       higher water concentration. (2)
d) 0.26 M (1)
e) i) Lower
   ii) 0.6 M sugar solution contains more sugar and
       less water than 0.26 M sugar solution. (2)
f) To prevent a second variable factor being
   introduced into the investigation. (1)
4 a) i) Compared with sea water, river water has a
         lower salt concentration.
   ii) When the tide is high, the salt concentration
       in the river estuary will increase.
   iii) When the tide is low, the salt concentration in
        the river estuary will decrease. (3)
   b) i) 00.00–06.00 and 13.00–19.00
      ii) Going out (2)
   c) i) 06.00–13.00 and 19.00–01.00
      ii) Coming in (2)
   d) To increase the reliability of the results. (1)

5 D (1)
6 a) In A, the HWC is outside and the LWC is inside.
      In B, the HWC is inside and the LWC is
      outside. (2)
   b) i) In A it will move up and in B it will move down.
      ii) In A, water molecules move along a
          concentration gradient from the dilute sugar
          solution (HWC) outside the Visking tubing
          to the concentrated sugar solution (LWC)
          inside. This increases the volume of liquid
          inside the Visking tubing and glass tube,
          making the level rise.
          In B, water molecules move along a
          concentration gradient from the dilute sugar
          solution (HWC) inside the Visking tubing
          to the concentrated sugar solution (LWC)
          outside. This reduces the volume of liquid
          inside the Visking tubing and glass tube,
          making the level drop. (4)
   c) It will increase. (1)
7 a) An inverse relationship: as the salt concentration
      increases, the water concentration decreases. (1)
   b) i) An inverse one: as the salt concentration
         increases, the number of pulsations decreases.
      ii) The higher the salt concentration of the
          bathing solution, the lower its water
          concentration and the lower the volume
          of water gained by osmosis by the animal
          immersed in it. (2)
   c) i) 1% salt solution
      ii) They worked at a faster rate. (2)
   d) i) Percentage concentration of salt in bathing
         solution
      ii) Temperature and pH of salt solution (3)
   e) To increase the reliability of the results. (1)
8 A, E, C, B, F, D (1)
9 a) Ion concentration is greater in the cell sap than in
      the pond water. (1)
   b) Potassium = 1200 : 1 and sodium = 71 : 1 (1)
   c) The data support the selective theory because
      they show that the plant accumulates different
      concentrations of different ions (rather than
      accumulating equal concentrations of them
      all). (1)
   d) i) The data dispute the suggestion.
      ii) In every case given in this example, ion uptake
          occurs from low to high concentration which
          is the reverse of diffusion. (2)

3 Producing new cells

1 a) W, Z, Y, X, U, V (1)
   b) U, X, Y and Z (1)
   c) A (1)
2 a) D (1)
   b) During mitosis and cell division, the number of cells doubles each time the process occurs but the diploid chromosome complement is maintained throughout. (2)
3 a) They can be grown in a nutrient solution to form a tissue culture. (1)
   b) 17 hours and 15 minutes (1)
   c) Increase in rate (1)
   d) 64 (1)
   e) i) B and C
      ii) C (2)
   f) i) Nutrition
      ii) Cells need energy to increase in size and build chromosomes before dividing. (2)

4 a) See table (2)

Stage	Time taken at 22 °C (hours)
P	5
Q	4
R	1
S	2

   b) i) 3:2
      ii) 16.67% (2)
   c) An increase in temperature <u>speeds up</u> the rate of mitosis and cell division. In this type of cell at 38 °C, stage Q would take <u>1 hour 20 minutes</u> and stage S would take <u>40 minutes</u>. (3)
   d) The rate of mitosis and cell division would slow down. (1)
5 B, F, C, D, A, E (1)
6 See table (6)

Error or omission in procedure	Correct procedure	Reason for following correct procedure
He did not heat the entire wire shaft of the loop to red heat.	Heat the loop's entire wire shaft to red heat.	To kill all microbes on the wire shaft.
He did not hold the lid of the Petri dish over the dish of nutrient agar.	Lift the lid of the Petri dish just enough to allow the inoculating loop to enter the dish.	To prevent spores of contaminants in the air from entering the dish.
He did not reflame the loop before using it for a second time.	Reflame the wire loop to red heat.	To kill any bacteria from the sour milk before taking a sample of fresh milk.

7 Apparatus B would need:
   - lime water not tap water
   - the same volume of liquid as apparatus A
   - sucrose not glucose solution
   - solution that had been boiled and cooled
   - 1% not 2% concentration
   - the length of its tube to be shortened (6)
8 a) i) Temperature
      ii) Three
      iii) As temperature increased, so did the volume of dough produced. (3)
   b) i) A's dough lacked sugar.
      ii) In B, the presence of sugar increased the volume of dough produced.
      iii) The yeast cells in B had plenty of sugar so they grew and respired more than those in A and released more CO_2, making the dough rise more. (3)
   c) 4 cm^3 (1)
   d) 1.5 times (1)
   e) i) Neither A nor B would show an increase in volume of dough.
      ii) The yeast would be destroyed at such a high temperature. (2)

4 DNA and the production of proteins

1 a) i) 3
      ii) 1
      iii) 2 (2)
   b) It is not long enough. A gene would be hundreds or thousands of bases in length. (1)

2   See Figure An 4.1 (1)

A
G
G
C
A
C

Figure An 4.1

3   a) 2800 (1)
    b) 20% (1)
4   B (1)
5   A (1)
6   C (1)
7   a) B, D, A, C (1)
    b) In the nucleus (1)
    c) Ribosome (1)
8   C (1)

5 Proteins and enzymes

1   a) Hydrogen peroxide (1)
    b) i) A and C
        ii) It brings about an increase in the rate of the reaction.
        iii) Oxygen – it relit a glowing splint. (4)
    c) B and D (1)
    d) A denatured enzyme's molecular structure has been irreversibly altered and therefore it can no longer fit its substrate. (2)
2   a) B (1)
    b) A is invalid because it contains $1\,cm^3$ not $2\,cm^3$ of pepsin and $2\,cm^3$ not $1\,cm^3$ of water. C is invalid because it contains boiled pepsin instead of fresh pepsin and acid instead of water. D is invalid because it contains boiled pepsin instead of fresh pepsin. (3)
3   a) Type of potential substrate (1)
    b) Volume of substrate, temperature of tubes and size of urease tablet (3)
    c) To mix the urease with the potential substrate. (1)
    d) i) B
        ii) The red litmus paper turned blue (indicating the release of ammonia gas formed by the breakdown of urea by urease). (2)

e) The specificity of an enzyme to its substrate (1)
4   a) See Figure An 5.1 (3)

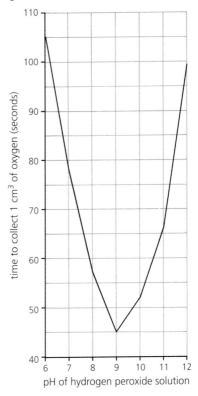

Figure An 5.1

    b) i) 9
        ii) 6 (2)
    c) 9 (1)
    d) Use a more detailed range of pH including values such as 8.5 and 9.5. (1)
5   a) It reduces the mass of food digested per hour. (1)
    b) 30% (1)
    c) They would fail to digest much of their food and therefore not absorb the food into the bloodstream. (1)
    d) 5:4:2 (1)
    e) 20% (1)
6   a) It would be digested to a soluble state. (1)
    b) 3 (1)
    c) 1 (1)
    d) i) 4
        ii) pH 3 at 30 °C
        iii) Pepsin's optimum pH is around 2.5 and its optimum temperature is around 40 °C. The values chosen for the answer to ii) are closest to these optimum values. (4)
    e) i) 2
        ii) It contains three variable factors.

iii) See Figure An 5.2 (4)

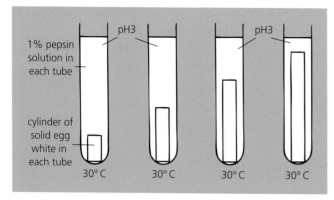

1% pepsin solution in each tube

pH3   pH3

cylinder of solid egg white in each tube

30°C   30°C   30°C   30°C

Figure An 5.2

7 a) Enzymes that control essential biochemical reactions in the body become denatured at this high temperature. The reactions slow down, with fatal consequences. (2)

b) The low pH destroys the attacking microbe's enzymes by denaturing them. This saves the food from being attacked and going bad. (2)

c) Any fungal colonies on the cheese grow more slowly in the refrigerator since enzyme molecules are less effective at combining with and acting on their substrate molecules at low temperatures. (2)

8 a) See Figure An 5.3 (3)

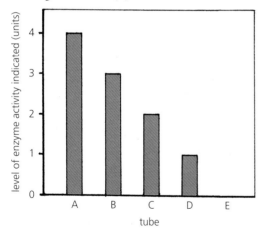

Figure An 5.3

b) i) This colour shows that no starch is present, thereby indicating that it has all been digested by the enzyme.

ii) This colour shows that much (or all) of the starch is present, thereby indicating that little (or none) of it has been digested by the enzyme. (2)

c) As concentration of aspirin increases, activity of amylase decreases. (1)

d) i) They differ in two ways – volume of aspirin 'solution' added and total volume of contents.

ii) An appropriate volume of water could be added to each of tubes A, B, C and D to make their contents equal in volume to the contents of tube E. (2)

6 Genetic engineering

1 a) D, F, B, E, C, A (1)

b) A plasmid from this type of bacterium would be cut open and the HGH gene inserted and sealed into it. The plasmid would be inserted into a bacterial host cell. (2)

c) See Figure An 6.1 (2)

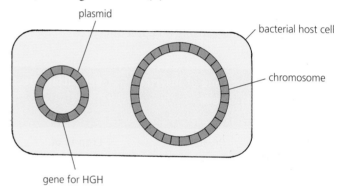

plasmid

bacterial host cell

chromosome

gene for HGH

Figure An 6.1

2 a) It is composed of a molecule of glucose and a molecule of galactose. (1)

b) The yeast cannot take in lactose through its membrane and it cannot break lactose down into glucose and galactose. (2)

c) It means the transfer of genetic material (as one or more pieces of chromosome) from one organism to another. (1)

d) It is able to make the enzyme lactose permease which allows lactose to enter the cell. It is also able to make lactase enzyme which breaks lactose down to glucose and galactose. (2)

e) D (1)

3 a) i) 82%

ii) 2005 (2)

b) The percentage of acres planted with GM crops is increasing with time. (1)

c) 8 times (1)

d) Soya bean (1)

e) i) 2000

ii) In that year, but in no other year, all three curves show a decrease in percentage of acres planted with GM crops, suggesting a shortage of seeds that year. (2)

4 a) Scientists inserted a gene into the bird's genome that produces a 'decoy' molecule that interferes with the virus' replication. (1)

   b) i) Yes

      ii) The transgenic chickens are unable to transfer the virus to other chickens. (2)

   c) i) Yes

      ii) Scientists have not found any differences to exist between the GM birds and their non-GM relatives. (2)

   d) There would be the economic benefit resulting from increased production of meat and eggs if all the birds remained free of avian flu. There would be a reduced risk of avian flu virus affecting humans and causing an epidemic. (2)

5 See table (5)

Genetically modified (GM) crop plant	Role of inserted gene	Beneficial effect
Rice	Production of chemical that can be converted in human body to vitamin A	**Crop is of improved nutritional value**
Apple	Blockage of production of chemicals that promote ripening	**Shelf life of fruit is extended**
Pea	**Production of a protein that acts as a natural insecticide**	Leaves able to resist attack by caterpillars
Soya	Production of a protein that gives resistance to weedkiller	**Crop survives but weeds die when weedkiller is applied**
Strawberry	**Production of a chemical that acts as a natural antifreeze**	Fruit protected against damage by frost

6 a) 1 = D, 2 = E, 3 = A, 4 = C, 5 = F, 6 = B (5)

   b) There is no correct answer to this question since it depends on the reader's own personal opinion.

7 a) A growth regulator gene from Pacific Chinook salmon and a promoter gene from ocean pout were inserted into the genetic material of Atlantic salmon. (2)

   b) i) The GM salmon has the potential to feed more efficiently and to survive for twice as long as its wild-type relative.

      ii) The GM salmon is a poorer swimmer and the male has a low level of reproductive success owing to its low sperm count. (4)

   c) If any did escape into the wild they would not be able to reproduce. (1)

   d) They are worried that people will resist buying the product thinking that something may be wrong with it because it has been produced by genetic engineering. (1)

7 Respiration

1 a) i) Corn oil, lard and olive oil

      ii) 38 kJ/g (3)

   b) i) Egg white

      ii) Sucrose

      iii) 19 kJ/g in both cases (4)

   c) 1:2:1 (1)

   d) 3 times (1)

2 See Figure An 7.1 (8)

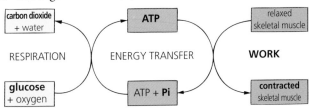

Figure An 7.1

3 a) 27 °C (1)

   b) 28 cm³/day (1)

   c) 39 °C (1)

   d) 3.5 times (1)

   e) 12 °C (1)

4 a) i) Q = 2, T = 18

      ii) R = ethanol, S = CO_2 (4)

   b) It will diffuse out of the cell. (1)

   c) i) 2

      ii) 38 (2)

5 a) i) It results in an increase in rate of ion uptake.

      ii) Some other factor (such as mass of sugar available for energy release) has run out.

      iii) It is an inverse relationship. Since sugar provides the energy for ion uptake, the number of units of sugar present in the cell sap decreases as the number of units of ion absorbed increases. (4)

   b) i) Approximately 37 °C

      ii) Enzymes needed for respiration (and energy release) are denatured at higher temperatures. (2)

6 B, D, A, C (1)

7 a) See table (8)

Athletic event	Volume of oxygen needed for event (l)	Volume of oxygen consumed during event (l)	Oxygen debt (l)	Percentage of energy obtained for event from aerobic respiration	Percentage of energy obtained for event from anaerobic respiration
100 metres	10	0.5	9.5	5	95
800 metres	25	8	17	32	**68**
1500 metres	36	18	18	**50**	50
10 000 metres	150	135	**15**	**90**	10
marathon (42 195 metres)	700	**686**	**14**	98	2

b) See Figure An 7.2 (4)

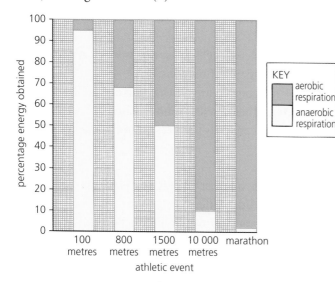

Figure An 7.2

8 Photosynthesis

1 a)
- make bell jar B the same size as A
- make B the same design of bell jar as A
- change the plant in B to geranium
- increase the volume of sodium hydroxide in B to the same as the volume of water in A
- change garden soil in B to potting compost
- reduce the quantity of soil (and size of plant pot) in B to equal that in A (6)

b) Temperature and light intensity (2)

2 a) $4 \, mm^3 \, CO_2/mm^2$ leaf (1)

b) 2 times (1)

3 a) Day 2, 12.00–15.00 (1)

b) Photosynthesis (1)

c) i) 05.00

ii) 21.00 (2)

d) i) Day 3

ii) The CO_2 curve shows the smallest dip suggesting least use for photosynthesis. (2)

4 a) i) 4 and 6

ii) 3 and 4

iii) 2 and 4

iv) In i), the only difference between the discs is the presence of light. In ii), the only difference between the discs is presence or absence of chlorophyll. In iii), the only difference between discs 2 and 4 is the presence or absence of CO_2 (which is absorbed by sodium hydroxide). (6)

b) To ensure that any starch found at the end of the experiment was produced during the experiment (and was not there before the experiment was begun). (1)

5 a) 2 (1)

b) Light (1)

c) D (1)

d) 4 (1)

6 a) Oxygen (1)

b) $0.04 \, cm^3/h$ (1)

c) i) Exchange the coolant for a stream of water heated to the required temperature.

ii) To allow the plant to become acclimatised to the new temperature. (2)

d) i) 55 °C

ii) The plant would probably be dead following the denaturation of its enzymes at this temperature. (2)

e) To ensure that CO_2 concentration is not a limiting factor when investigating the effect of varying temperature. (1)

f) Keep the temperature constant (at, say, 20 °C) and then vary the light intensity by varying the distance of the light source from the plant or by using a lamp with a dimmer switch. (2)

7 a) i) Sugar production as g/kg of dry plant
   ii) Time (taken by the plant to produce known masses of sugar)
   iii) Rate of oxygen production (3)
   b) 0.6% (1)
   c) See Figure An 8.1 (3)

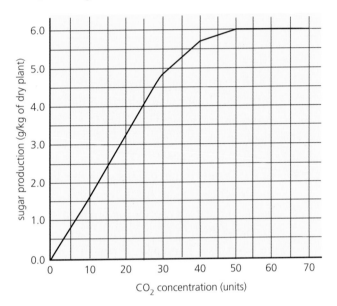

Figure An 8.1

d) 2.5 g/kg of dry plant (1)
e) i) Each increase in CO_2 concentration between 0 and 50 units was accompanied by a rise in the graph, indicating an increase in sugar production (in other words, photosynthesis) and showing that CO_2 concentration was holding up the process until 50 units.
   ii) Each increase in CO_2 concentration from 50 units onwards resulted in a continuous levelling off of the graph, indicating no further increase in sugar production despite the presence of plenty of CO_2. Therefore CO_2 concentration was no longer limiting the process.
   iii) Light intensity (5)
8 a) i) To make the experiment valid by preventing the introduction of a further variable factor.
   ii) Temperature and volume of water added to soil (3)
   b) 3.4 kg/plant (1)
   c) 4.9 (1)

d) i) 6.7 kg/plant
   ii) 10.1 (2)
e) Increased light intensity (1)
f) To make the results reliable. (1)
g) To act as the control. (1)

9 Cells, tissues and organs

1 a) Sperm (1)
   b) Its tail enables it to swim to the egg. Its mitochondria release energy (during aerobic respiration) for movement. (2)
2 a) Root hair; absorption of water (2)
   b) Sieve tubes and companion cells: sieve tubes lack nuclei but have sieve-like end plates; companion cells have nuclei but lack sieve-like end plates. (4)
   c) i) Yes
      ii) Different cells are specialised to carry out particular functions. (2)
3 a) It has columnar epithelial cells and goblet cells. (2)
   b) Its epithelial cells are not ciliated. It has enzyme-producing cells not present in the trachea. (2)
4 a) B (1)
   b) D (1)
5 a) It is the smallest unit that can lead an independent life. (1)
   b) Respiration, growth, reproduction and feeding (4)
   c) A unicellular organism's body consists of one cell whereas that of a multicellular organism consists of more than one cell. (1)
   d) i) 2×10^{12}
      ii) 30 times (2)
   e) Its biconcave shape presents a large surface area through which oxygen can pass into the cell. (2)
   f) Tissue (1)
   g) i) Organ
      ii) It is composed of many different tissues. (2)
   h) Cell, tissue, organ, human body (1)
6 1 = molecule, 2 = organelle, 3 = cell, 4 = tissue, 5 = organ, 6 = system, 7 = organism (6)

10 Stem cells and meristems

1 a) E, B, D, A, C (1)
   b) So that the type of skin used is as similar as possible to the damaged skin. (1)
   c) This type of graft uses a smaller sample of skin and involves less healing time (and causes less scarring). (2)
   d) Because it is the injured person's own cells that are being used. (1)

2 a) See table (2)

Region of spinal cord injured	Number of cases	Number of cases showing improvement	Percentage number of cases showing improvement	Number of cases not showing improvement	Percentage number of cases not showing improvement
Cervical	220	165	75	55	25
Thoracic	180	108	60	72	40
Lumbar	300	159	53	141	47

b) See Figure An 10.1 (4)

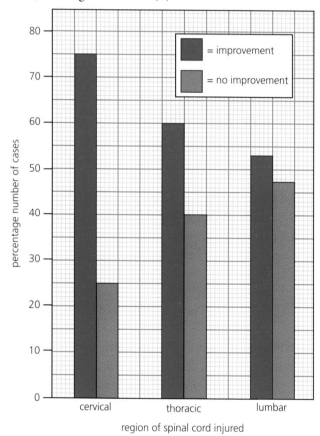

Figure An 10.1

c) 3 times (1)

d) The number of cases showing improvement seems to decrease with increasing distance from the brain. (1)

e) Study a similar group of people who had not received stem cell treatment. (1)

3 a) X = 16.5, Y = 5.5 (2)

b) See Figure An 10.2 (5)

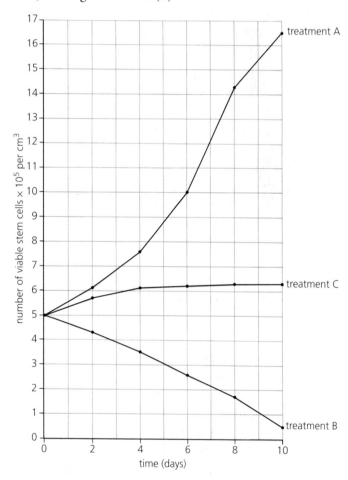

Figure An 10.2

c) Treatment A brings about an increase in number of stem cells. Treatment B brings about a decrease in number of stem cells. (2)

d) Control (1)

e) To increase the reliability of the investigation's results. (1)

f) Temperature, pH and chemical composition of growth medium (3)

g) i) 230%
   ii) 90% (2)

4 a) A = phloem tissue, B = xylem vessel, C = meristematic cell, D = root cap, E = epidermal cell (4)

   b) i) B and E
      ii) C (3)

5 D (1)

6 a) A (1)
   b) B (1)

7 a) For additional support and for transport of water to new tissues. (2)

   b) i) Winter
      ii) The temperature is too low for rapid growth and division of cells to occur. (2)

   c) i) 5 years
      ii) By counting the rings (2)

   d) D (1)

8 a) See Figure An 10.3 (3)

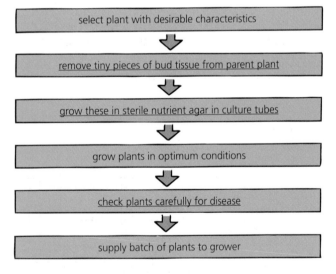

Figure An 10.3

b) Light intensity and temperature (2)

c) Sterile nutrient agar is used to grow the plants. (1)

d) 'Fruit that all ripen at the same time' (1)

e) Plant cloning (1)

f) i) A new disease-causing organism

   ii) The members of the crop population make up a clone and are genetically identical to one another. If one catches the disease, they will all become infected. (2)

11 Control and communication

1 a) Side 2 (1)

   b) A = cerebrum, B = medulla, C = spinal cord, D = cerebellum (4)

   c) i) T
      ii) T
      iii) F – cerebrum
      iv) F – cerebellum
      v) T
      vi) F – medulla (6)

2 a) i) 10
      ii) 11
      iii) 4
      iv) 11 (4)

   b) 2 and 4 (2)

   c) So that the person being tested could not anticipate in advance which test (heat or cold) was about to be applied. (1)

   d) To ensure that they did not see which test (heat or cold) was being applied. (1)

   e) To allow time for the previous sensation of heat or cold to fade before applying the next test. (1)

   f) i) More locations on that person's body would need to be tested.
      ii) Many more people would need to be tested. (2)

   g) The blunt pin could be replaced by a sharper one. (1)

   h) It acts as a warning to the person that the affected tooth is in need of attention. (1)

3 a) Rate of conduction of nerve impulses is much faster in warm-blooded than in cold-blooded animals. Rate of conduction of nerve impulses is much faster in giant nerve fibres than in normal ones in the same type of animal. (2)

   b) It allows it to respond quickly to danger and move away rapidly. (1)

   c) 20 times (1)

4 See table (11)

Reflex action	Stimulus	Response	Protective function	Can be altered partly by voluntary means?
Blinking	**Object touching eye**	Contraction of eyelid muscle	**Helps prevent damage to eye**	Yes
Peristalsis	Presence of food in gut	**Contraction of muscles in gut wall**	Ensures movement and therefore efficient digestion of food	No
Sneezing	Foreign particles in nasal tract	Sudden contraction of chest muscles	**Unwanted particles removed from nose**	Yes
Dilation of pupil	**Dim light**	**Movement of iris muscle**	**Vision in dim light improved, reducing the risk of an accident**	No

5 a) Follicle-stimulating hormone (FSH) and luteinising hormone (LH) (2)
  b) Oestrogen and progesterone (2)
  c) i) 1 = progesterone, 2 = LH, 3 = FSH, 4 = oestrogen
     ii) LH (5)
6 a) A higher percentage of people are affected with diabetes in late middle and old age than among younger or very old people. More men than women are diabetic. (2)
  b) Children under the age of 15 have not been included in the data. (1)
7 a) 44 units (1)
  b) i) 220 mg/100 ml
     ii) 460 mg/100 ml (2)
  c) i) 2 hours
     ii) In the liver (where it would be stored as glycogen). (2)
  d) Graph 1 shows that after ingesting glucose, Y's insulin concentration failed to increase. Graph 2 shows that after ingesting glucose, Y's glucose concentration remained high and failed to return to its starting level within the 5 hours of the test. (2)

12 Reproduction

1 a) 0.06 mm (1)
  b) i) X = 21 million/ml, Y = 19 million/ml, Z = 26 million/ml
     ii) Y (4)
  c) If many sperm are present then there is an increased chance that at least one of them will successfully reach an egg and fertilise it. (1)

2 See table (7)

Site of:	Number of structure	Name of structure
Copulation	4	Vagina
Egg production	1	Ovary
Fertilisation	2	Oviduct
Embryo development	3	Uterus

3 a) 4 (1)
  b) B (1)
4 a) See Figure An 12.1 (3)

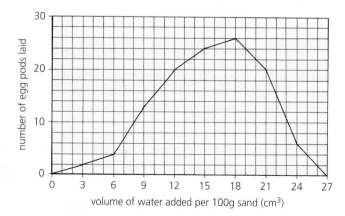

Figure An 12.1

  b) i) Volume of water added to sand
     ii) 18 cm³ water per 100 g sand (2)
  c) i) 10
     ii) 13.5 cm³ and 20 cm³ (3)

d) To increase the reliability of the results. (1)

e) To prevent them drying out and to prevent them from being eaten by predators. (2)

5 a) X = 400, Y = 5, Z = 10 (3)

   b) i) An inverse relationship (the smaller the number of eggs produced annually, the longer the length of parental care).

      ii) A direct relationship (the longer the length of parental care, the higher the percentage of young that survive). (2)

   c) It is very difficult, if not impossible, to collect accurate data for some of the values (such as exactly how many young cod out of 4 million survive after a year). (1)

6 a) A = *Viola canina*, B = *Viola lutea* (2)

   b) Flowers sweetly scented, leaf stalks hairy, style extended into hook-like stigma (and style not expanded into ball-like stigma) (3)

7 a) A (1)

   b) D (1)

8 a) i) 1 and 3

      ii) 5 and 7 (2)

   b) B (1)

   c) Type of pollen grain (1)

   d) They differ in two ways (type of sugar and concentration of sugar). (1)

   e) Temperature and light intensity (2)

   f) Repeat the experiment several times (or set up several replicas of each condition). (1)

9 See table (7)

13 Variation and inheritance

1 a) i) 2.1 to 3.3 g

      ii) 2.6 g (2)

   b) 11 (1)

   c) 3.3 g (1)

   d) 10 (1)

   e) 80 (1)

   f) 10% (1)

   g) 35 (1)

   h) 2.7 g (1)

2 a) Continuous (1)

   b) See Figure An 13.1 (4)

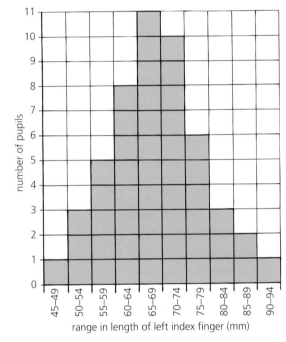

Figure An 13.1

Insect-pollinated flower		Wind-pollinated flower	
Structural feature	**Reason**	**Structural feature**	**Reason**
Flower large with brightly coloured petals, scent and nectar	To attract insects which will eat nectar and collect pollen	**Flower small, lacking bright colour, scent and nectar**	No visit from insect required
Anthers firmly attached inside flower	To be in a position where insects are likely to brush against them	Anthers loosely attached and hanging out of flower	**To enable them to be shaken and to allow pollen to be carried away by wind**
Pollen grains sticky or with rough surface	**To enable them to stick to insect's body easily**	Pollen grains light and smooth	**To enable them to be carried in air currents without sticking together**
Stigmas inside the flower with sticky surface	**To be in a position where insect is likely to brush against them and leave pollen stuck to them**	Stigmas hanging out of flower and feathery in structure	**To be in a good position and present a large surface area for trapping pollen**

c) 10 (1)

d) i) 65–69 mm

ii) 11 (2)

e) 12 (1)

3 a) See Figure An 13.2 (4)

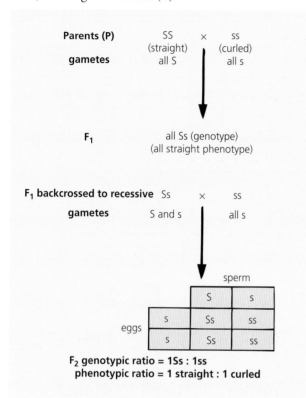

Figure An 13.2

b) The allele for curled wing is recessive and masked by straight wing, the dominant allele. (1)

c) i) 1:1

ii) 84 straight to 84 curled

iii) Fertilisation is a random process that involves an element of chance. (4)

4 a) Gg and gg (1)

b) i) Gg and gg

ii) 1:1 ratio (2)

c) GG (1)

5 a) From her father (1)

b) From her paternal grandmother and maternal grandfather (via both of her parents) (2)

c) See Figure An 13.3 (8)

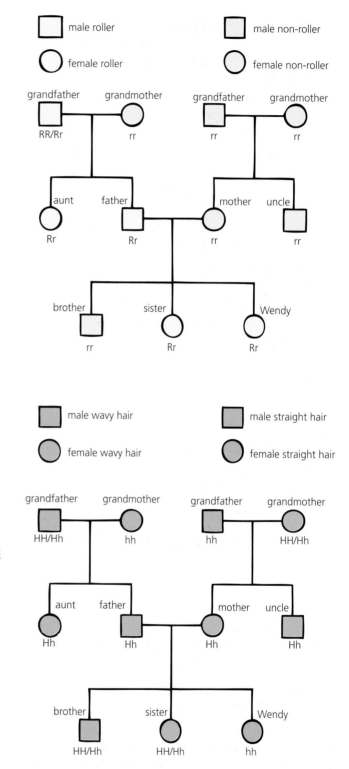

Figure An 13.3

6 **a)** and **b)** See Figure An 13.4 (4 + 4)

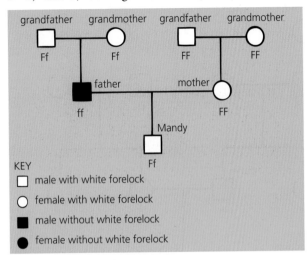

KEY
□ male with white forelock
○ female with white forelock
■ male without white forelock
● female without white forelock

Figure An 13.4

c) She has a white forelock. (1)

7 **a)** See Figure An 13.5 (4)

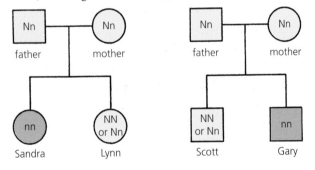

Figure An 13.5

b) **i)** A and B

ii) 1 in 4

iii) See Figure An 13.6 (5)

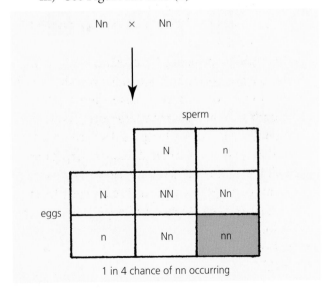

1 in 4 chance of nn occurring

Figure An 13.6

8 **a)** See Figure An 13.7 (2)

		pollen							
		■■■	■■□	■□■	□■■	■□□	□■□	□□■	□□□
ovules	■■■	6	5	5	5	4	4	4	3
	■■□	5	4	4	4	3	3	3	2
	■□■	5	4	4	4	3	3	3	2
	□■■	5	4	4	4	3	3	3	2
	■□□	4	3	3	3	2	2	2	1
	□■□	4	3	3	3	2	2	2	1
	□□■	4	3	3	3	2	2	2	1
	□□□	3	2	2	2	1	1	1	0

Figure An 13.7

b) **i)** Six shades of red (+ white)

ii) Shade 3 (three red alleles)

iii) Shade 6 (six red alleles)

iv) 9.38% (4)

c) 4:3 (1)

14 Transport systems in plants

1 **a)** D (1)

b) B (1)

c) A (1)

d) C (1)

2 C (1)

3 **a)** Pink to blue (1)

b) **i)** The paper was exposed to the air in the room for some time on its way to the leaf. Therefore it could have picked up water vapour from the air in the room.

ii) Keep the desiccator as close as possible to the plant to be used (2)

c) An exact repeat of the procedure should have been followed except that the cobalt chloride paper should have been enclosed in a space formed by a slide, Plasticine and a non-living surface such as a second glass slide or a dead leaf. (2)

4 **a)** Q = 259.32, R = 24.48 (2)

b) 0.51 g/h (1)

c) More water is taken in by the plant than is given back out again because the plant retains a little for photosynthesis (and maintenance of turgor). (1)

d) i) The rate of water loss will be dramatically reduced.
  ii) Stomata close in darkness preventing transpiration. (2)
5 a) None of its stomata were blocked with Vaseline. (1)
  b) All of its stomata were blocked with Vaseline. (1)
  c) i) X = 0.99, Y = 3.04%
    ii) Leaf 3
    iii) Most of the stomata are on the lower side of this type of leaf. (4)
  d) i) The leaf with its lower side coated with Vaseline.
    ii) They are on the upper surface of the leaf.
    iii) A water lily leaf floats on water so it needs to have its stomata on its upper surface to allow CO_2 in the air to enter its air spaces and be available for use in photosynthesis. (4)
6 a) i) Xylem
    ii) Phloem (2)
  b) The leaves in shoot P have become wilted because they have lost water by transpiration but have been unable to replace it since their xylem vessels are blocked. The leaves in shoot Q have remained turgid because their xylem vessels are not blocked and they are able to replace any water losses. (2)
7 a) i) Water (and mineral salts)
    ii) Water is transported in xylem and some would now escape via the stylet. (2)
  b) i) Sugar solution
    ii) Sugar solution is transported in phloem and some would now escape via the stylet. (2)
  c) i) Up
    ii) In this case it would be moving up to feed the flower. (2)
8 a) Its leaves lack green chlorophyll, which is essential to trap light for photosynthesis. (1)
  b) A = dodder, B = gorse (1)
  c) X = xylem, Y = phloem (2)
  d) It absorbs it from the phloem tissue of the gorse plant. (1)
  e) It absorbs them from the xylem tissue of the gorse plant. (1)
  f) Dodder is the parasite and gorse is the host. (1)

15 Animal transport and exchange systems

1 a) i) Unicellular animal
    ii) Frog
    iii) Lungs (3)
  b) Gills (1)

2 a) i) 1 = artery, 2 = capillary, 3 = vein, 4 = heart
    ii) 3 and 4 (6)
  b) i) Food and oxygen
    ii) Carbon dioxide (3)
3 a) See table (5)

Part of body	Rate of blood flow (cm³/min)		
	At rest	Light exercise	Strenuous exercise
Heart muscle	**250**	350	750
Brain	750	750	**750**
Kidneys	**1100**	900	600
Skin	500	**1500**	1900
Gut	1400	1100	**600**
Skeletal muscle	1200	**4500**	**12 500**

  b) x-axis = part of body; y-axis = rate of blood flow (cm³/minute) (2)
  c) i) It makes rate of blood flow increase.
    ii) This supplies the muscles with the increased quantity of food and oxygen needed to generate extra energy for strenuous exercise. (2)
  d) i) Skin and heart muscle
    ii) Kidneys (2)
  e) i) Brain
    ii) The brain is the body's control centre. To be reliable and stay in perfect working order, it must remain unaffected by exercise. (2)
  f) i) It would become redder.
    ii) It would be affected by an increased rate of blood flow. (2)
4 a) 1 = high, 2 = alveolus, 3 = capillary, 4 = oxyhaemoglobin, 5 = red, 6 = blood, 7 = cell, 8 = low, 9 = oxygen, 10 = haemoglobin (9)
  b) Blood in the pulmonary arteries (on its way from the heart to the lungs) is deoxygenated and lacking in oxyhaemoglobin, making it dark red. Blood in the pulmonary veins (on its way from the lungs to the heart) is oxygenated and rich in oxyhaemoglobin, making it bright red in colour. (2)
5 598:1 (1)
6 a) B (1)
  b) C (1)
  c) D (1)

7 a) For parts **i)**, **ii)** and **iii)** see Figure An 15.1 (6)

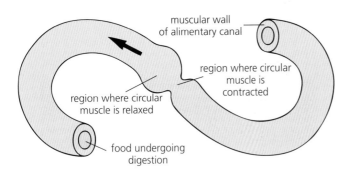

Figure An 15.1

b) i) Small intestine

   ii) The tube is too narrow to be the large intestine and it contains too many bends to be the oesophagus. (2)

c) Oesophagus and large intestine (2)

8 a) X = 300, Y = 200 (2)

b) Some would be used for bone formation and therefore less would pass out of her body in wastes. (1)

9 a) Peristalsis (1)

b) 125 minutes (1)

c) i) 2, 3, 11

   ii) See Figure An 15.2 (4)

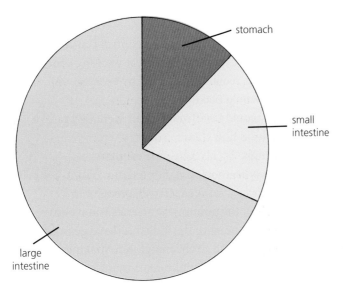

Figure An 15.2

d) V = 201, W = 755, X = 160, Y = 614, Z = 692 (5)

16 Effects of lifestyle choices

1 a) 2012 group (1)

b) Leisure activities such as computer games and television viewing made many of the 2012 teenagers less active than those of 1962. Therefore some of the 2012 teenagers were not burning up all the energy contained in their food, leading to weight gain. (2)

c) Compared with teenagers in 1962, those in 2012 were often eating food rich in sugar or fat such as soft drinks, crisps and chocolate. This made the 2012 teenagers more likely to gain weight. (1)

d) Drinking a can of sugary soft drink means taking in extra sugar which the body may not need. Eating an orange is better because it is a natural food which contains several useful nutrients in addition to vitamin C. (1)

e) The person could eat more fruit in place of chocolate and crisps. They could play more sport instead of watching television. They could walk to school sometimes instead of always taking the bus or being driven there. (3)

f) To make the results more reliable. (1)

2 a) *Physical effects*: constipation, diarrhoea, headache, indigestion, itchy skin
Mental effects: anxiety, depression, inability to cope, inability to show feelings, short temper (2)

b) Approach their employer in order to try to reduce stress at work or consider changing jobs. (1)

3 a) i) 6 g

   ii) X = 5.8 g, Y = 6.3 g, Z = 5.7 g

   iii) Y (5)

b) i) 9

   ii) 5 (2)

4 a) C, A, D, B (1)

b) i) It involves several variable factors (instead of just one).

   ii) It could be made valid by getting all the students to behave in the same way as one another during the 24 hours before the experiment (for example, all eating similar food at the same times, all going to bed at the same time etc.). (3)

5 a) i) B

   ii) C (2)

b) i) D

   ii) B (2)

c) D (1)

6 a) See Figure An 16.1 (4)

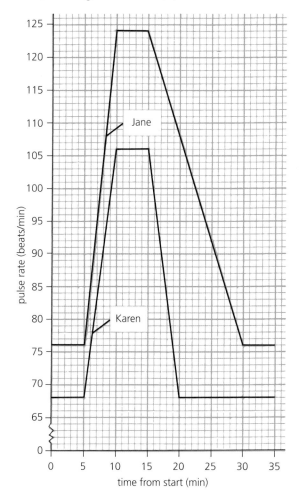

Figure An 16.1

b) 76 beats/min (1)
c) 124 beats/min (1)
d) 15 minutes (1)
e) 68 beats/min (1)
f) 106 beats/min (1)
g) 5 minutes (1)
h) i) Karen
   ii) Jane (2)
7 a) See table (3)

Country	Number of people per 100 000 of population dying of heart disease each year	
	Women	Men
Japan	10	52
Germany	64	246
Scotland	140	510
France	35	115

b) 80 (1)
c) 395 (1)
d) i) 62
   ii) 310
   iii) 5 times (3)
e) 14:1 (1)
8 a) i) Female
      ii) Male (2)
   b) An increase (1)
   c) i) A decrease
      ii) Rate is a *relative* measure. Although there is a higher rate of incidence among 80–84 year olds per 100 000 compared with 60–64 year olds, there are far fewer men of age 80–84 in the population, therefore the *actual* number of cases is much lower. (3)
9 a) D (1)
   b) It is the only one in support of privatisation of health care. (1)

17 Biodiversity and the distribution of life

1 a) A (1)
   b) D (1)
   c) F (1)
   d) 4.6 (1)
   e) 7.6 (1)
   f) C (1)
   g) 3 (1)
   h) C (1)
2 a) As latitude decreases, the number of species of flowering plants and snails increases. (1)
   b) A decrease in latitude is accompanied by a decrease in distance from the equator, the warmest region on Earth. The increase in temperature provides optimum growing conditions for an increased number of species. (2)
3 a) A (1)
   b) A (1)
   c) D (1)
4 a) P (1)
   b) W (1)
   c) Frequent (1)
   d) Oil (1)
   e) i) Q
      ii) They are closest to the oil. (2)

f) i) W

   ii) The crabs at site W were furthest away from the source of pollution. The crabs at site W were less affected by the oil because the flow of river water towards the sea helped to prevent the oil from reaching them. (3)

5 a) They are unable to digest the fur and bones of their prey. (1)

   b) See Figure An 17.1 (3)

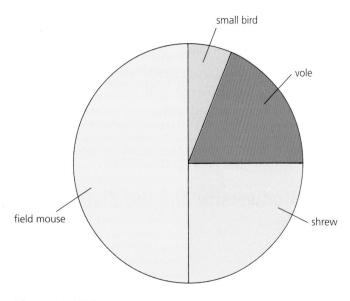

Figure An 17.1

c) wheat ⟶ field mouse ⟶ barn owl (1)

d) i) No

   ii) There are no poultry remains in the pellets, mainly the remains of small mammals. (2)

e) The owls help to keep their crops clear of mice. (1)

6 a) 4 (1)

   b) 4 (1)

   c) 5 times (1)

   d) 23 (1)

   e) i) Many prey organisms were being eaten by predators.

   ii) The predators reproduced and the presence of the plentiful supply of food enabled them and their offspring to survive. (2)

7 a) See Figure An 17.2 (5)

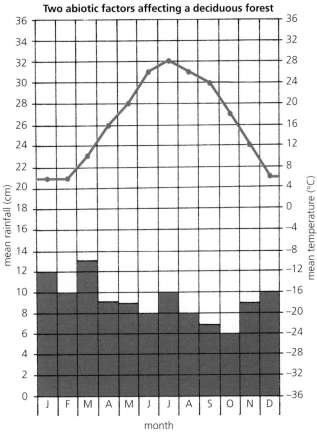

Figure An 17.2

b) The mean rainfall for deciduous forest is lower than that for tropical rainforest.
   The mean rainfall for deciduous forest is fairly steady whereas that for tropical rainforest is much more variable.
   The mean temperature for deciduous forest varies from low in winter to high in summer whereas the temperature of tropical rainforest remains fairly steady throughout the year. (3)

c) A (1)

8 a) i) 26 tonnes/hectare/year

   ii) 16 tonnes/hectare/year (2)

   b) Temperate forest (1)

   c) i) Tropical rainforest

   ii) Desert (2)

   d) Rainfall (1)

9 See table (14)

Organism	Habitat	Niche
Bracken	Open countryside	Producer that rapidly spreads by growth of underground stems; dominates environment by choking out rivals; produces poison that deters grazing animals
Alder plant	Exposed river banks	Producer whose roots resist underwater rot and survive in water-logged soil; able to exploit environment unavailable to other trees
Red deer	Woodland and moorland	Herbivorous consumer showing population increase in absence of its extinct natural predator (wolf)
Common seal	Salt water and sand banks	Fish-eating consumer with few serious rivals; temporary drop in numbers in recent years caused by viral disease
Brown trout	Fresh water	Insect-eating consumer suffering intense competition from introduced rainbow trout
Mole	Underground burrow	Nocturnal worm-eating consumer preyed on by owls

10 a) i) 3 = frequent
   ii) 5 = abundant
   iii) Between 2 and 2.5 metres (3)
  b) i) Star
   ii) The vacated part of the slope was up near the high tide mark and would not be as wet as lower down the slope. Perhaps the conditions were not wet enough there for acorn barnacles. (2)
  c) i) Acorn
   ii) The star barnacles (which are more delicate) no longer had to compete with the sturdier acorn barnacles, so they could colonise the vacated territory. (2)

18 Energy in ecosystems

1 a) Z (1)
  b) i) X
   ii) It is the most numerous. It is unable to survive in deep water where there is too little light for photosynthesis. (3)
  c) Y is the primary consumer because it is more numerous than Z. Z is the secondary consumer because it is less numerous than Y. (2)

2 a) i) oak tree ⟶ woodmouse ⟶ owl
   ii) plant plankton ⟶ crustacean ⟶ salmon ⟶ fish louse
   iii) algae ⟶ animal plankton ⟶ anchovy ⟶ tuna
   iv) heather ⟶ mountain hare ⟶ eagle (4)

  b) i) = natural woodland
   ii) = sea loch
   iii) = ocean
   iv) = moorland (3)
  c) i) = C
   ii) = B
   iii) = A
   iv) = D (3)

3 a) i) 100 kg
   ii) 10 kg (2)
  b) i) X = sheep and Y = salmon. Herbivores are less efficient at both absorbing energy from food and converting energy into body tissues. Therefore the two lower figures in the table refer to the sheep (a herbivore) and the two higher figures refer to the salmon (a carnivore).
   ii) It is used for movement and lost in waste materials. (4)
  c) i) The animals are warmer indoors and therefore need less of their energy reserves to maintain their body temperature. Instead of being used to generate heat, the energy is built into body tissues.
   ii) The animals cannot move about and therefore use less energy. Instead of being used for movement, energy reserves are built into body tissues. (4)

d) i) Trout fish farm

ii) Unlike a mammal, such as deer, a fish such as trout is ectothermic ('cold-blooded') and does not use its food reserves to keep warm. Therefore more energy is available to be built up into the fish's body tissues. (2)

4 a) See Figure An 18.1 (2)

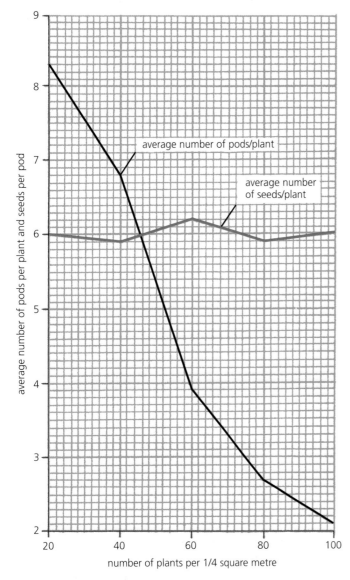

Figure An 18.1

b) i) Average number of pods per plant

ii) Decreases

iii) Water and soil minerals (4)

c) i) 996

ii) 1260 (2)

d) i) 100

ii) A higher number of seeds is produced. (2)

e) Repeat the experiment and see if similar results are obtained. (1)

5 See Figure An 18.2 (7)

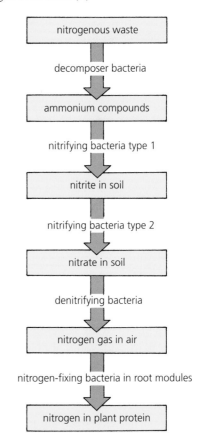

Figure An 18.2

6 a) Different numbers of seeds were planted. (1)

b) Size of pots, mass of cotton wool and volume of water added (3)

c) See table (2)

Number of seeds planted	Number of healthy seedlings with green leaves	% number of healthy seedlings with green leaves
100	87	87
200	178	89
300	204	68
400	228	57
500	225	45

d) See Figure An 18.3 (3)

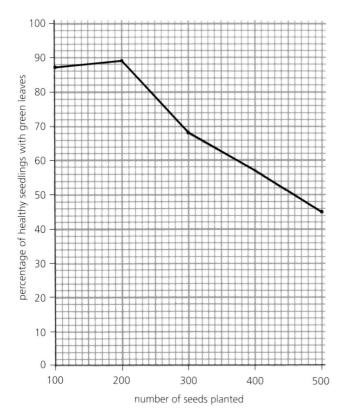

Figure An 18.3

e) As competition increases, the percentage number of healthy seedlings in the population decreases. (1)

f) Light (1)

g) See Figure An 18.4 (2)

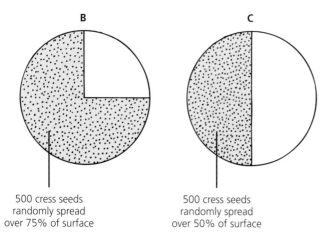

B

C

500 cress seeds randomly spread over 75% of surface

500 cress seeds randomly spread over 50% of surface

Figure An 18.4

7 a) Presence of food, oxygen and favourable temperature (3)

b) Food (1)

c) *P. caudatum* = 220 in 0.5 ml of medium; *P. aurelia* = 470 in 0.5 ml of medium (2)

d) Day 6 (1)

e) i) Yes

ii) When it was cultured with *P. caudatum*, its number did not reach as high a level as it did when it was cultured alone. (2)

f) i) Day 17 or 18

ii) Interspecific (2)

8 a) $20–120\,m^2$ (1)

b) The food supply is insufficient (therefore costs would outweigh benefits). (1)

c) When the energy expended covering large distances is greater than the energy gained from the food obtained. (1)

d) i) $70\,m^2$

ii) This gives the greatest net energy gain. (2)

19 Sampling techniques and measurements

1 a) $100\,m^2$ (1)

b) 600 (1)

c) 600 (1)

d) i) No (1)

ii) The number of dandelion plants remained unchanged. (2)

e) i) Too few quadrats were used.

ii) Increase the number of quadrats in the sampling process. (2)

2 a) The organisms move down and away from hot, bright conditions. They pass through the muslin and gather at the bottom of the funnel. (2)

b) A drop of water from the glass funnel could be added to a slide and covered with a cover slip ready for viewing. (1)

c) It creates dry conditions which would kill aquatic organisms. (1)

3 a) C (1)

b) D (1)

4 a) GH (1)

b) Yellow (1)

c) No (1)

d) The use of quadrats (1)

5 **a)** See Figure An 19.1 (4)

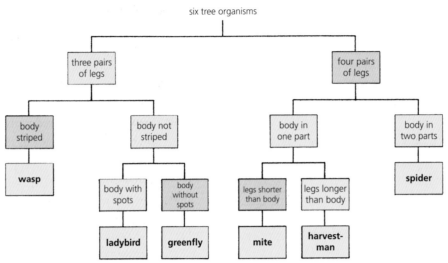

six tree organisms

three pairs of legs

four pairs of legs

body striped

body not striped

body in one part

body in two parts

wasp

body with spots

body without spots

legs shorter than body

legs longer than body

spider

ladybird

greenfly

mite

harvest-man

Figure An 19.1

b) A = greenfly, B = spider, C = ladybird,
    D = harvestman, E = wasp, F = mite (1)
c) **1** three pairs of legs .... go to **2**
        four pairs of legs .... go to **4**
    2 body striped........................... **wasp**
        body not striped .... go to **3**
    3 body with spots .... **ladybird**
        body without spots................ **greenfly**
    4 body in one part..................... go to **5**
        body in two parts.................... **spider**
    5 legs shorter than body .... **mite**
        legs longer than body .... **harvestman** (3)
6 a) Light intensity (1)
  b) It decreases from bright to fairly dim. (1)
  c) i) It changes from abundant to absent.
    ii) A direct relationship: as one increases, so does
        the other.
   iii) Meadow buttercups are plants that need
        bright light for photosynthesis. (3)
  d) i) They only took one light intensity reading at
        each quadrat.
    ii) They could take several light meter readings at
        each of the ten locations along the transect to
        increase the reliability of the results. (2)
  e) The dampness of the soil and the pH of the
    soil (2)

7 **a)** See Figure An 19.2 (4)

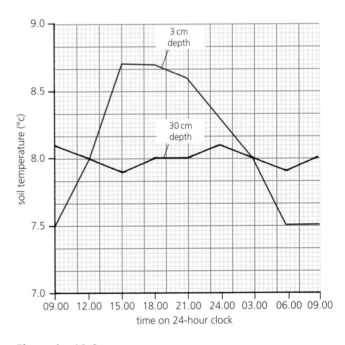

Figure An 19.2

  b) Every 3 hours (1)
  c) A thermometer (or a temperature probe) (1)
  d) 12.00 and 03.00 (1)
  e) It is near the surface and therefore it is affected
    by heat from the Sun. (1)

f) It is near the surface and is therefore affected by the drop in temperature that occurs above ground during the night. (1)

g) i) Soil at 30 cm
   ii) It is located so deep down that it is unaffected by temperature changes at the soil surface. (2)

h) i) It would show less variation.
   ii) The leaves would shield it from the Sun's rays and it would not warm up as much as exposed soil. (2)

8 a) A (1)
  b) B (1)
  c) B (1)

9 a) 0–10 cm/second (1)
  b) 90–100 cm/second (1)
  c) 30–40 and 40–50 cm/second (1)
  d) The number of A decreases as the water speed increases. The number of B increases as the water speed increases. (2)
  e) i) B
     ii) Coarse gravel is moved by a water speed of 75 cm/second. At this speed many more B larvae than A larvae are found in the river water. (2)

20 Adaptation, natural selection and evolution

1 a) See Figure An 20.1 (3)

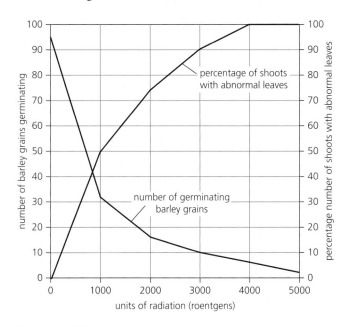

Figure An 20.1

b) As the number of units of radiation increases, the number of barley grains that germinate decreases. (1)

c) To make the results reliable (1)

d) See table (2)

Units of radiation (roentgens)	0	1000	2000	3000	4000	5000
Percentage number of shoots with abnormal leaves	0	50	75	90	100	100

e) See Figure An 20.1 (2)

f) As the number of units of radiation increases, the percentage of shoots with abnormal leaves increases. (1)

g) 1750 roentgens (1)

2 a) X = 17.2, Y = 17.0 (2)
  b) i) Female
     ii) Wales (2)
  c) i) 1038
     ii) 1104 (2)

3 a) Unlike a kangaroo rat, a camel cannot retreat into a cool, underground burrow during the day. (1)
  b) Sweating (1)
  c) i) No. The liquid in the stomach is not pure water. It contains fermented food.
     ii) Yes. The liquid is equal in water concentration to blood so it would help to relieve dehydration. (4)
  d) i) No. Any water gained from fat metabolism is lost during the increased breathing needed to absorb extra oxygen.
     ii) As an energy store. (3)
  e) Yes. It insulates the camel against the desert heat, keeping it cool and reducing its need to sweat. (2)
  f) Its blood does not become thick and sticky so its heart continues to pump the blood round the body as efficiently as normal. (1)

4 Despite *over-production* of offspring, a population explosion does not occur because the environment does not provide enough food. *Competition* for this and other limited resources occurs. Since *variation* exists among the members of the population, *natural selection* will favour those that are better adapted in some way (faster, stronger, etc.). They will survive and the weaker animals will lose out in the struggle and die. (4)

5 a) An inverse relationship: the percentage number of melanic moths decreases as the number of lichen species increases. (1)
  b) Melanic (dark) moths lose their selective advantage in clean places outside the city where

they are no longer camouflaged and are therefore easily spotted by predators. (2)

c) The line graph for pale moths would have shown a trend similar to that of lichens. (1)

d) The polluted area became cleaner as a result of the Clean Air Acts. Therefore the melanic moths had begun to lose their selective advantage. (2)

e) Abiotic (1)

6 a) Bacteria resistant to erythromycin enjoyed a selective advantage and became more common. Therefore fewer and fewer cases responded to treatment over the years. (1)

b) Very few or no cases would have been successfully treated by erythromycin. (1)

c) i) They prescribed a different antibiotic.

ii) Some mutant bacteria were already resistant to this second antibiotic. (2)

7 a) They kill or prevent growth of many bacteria in the mouth, thereby reducing tooth decay. (1)

b) Bacteria resistant to the antibiotic will be selected. They will multiply, fail to respond to the antibiotic and cause tooth decay. (1)

8 a) Field mice arrived with humans on the islands a very long time ago. They have remained isolated ever since and have begun to take their own course of evolution. (2)

b) i) If complete isolation continues

ii) If isolation is interrupted by human activities that enable mainland mice to reach the islands and breed with the island mice. (2)

21 Human impact on environment

1 See table (5)

2 a) See Figure An 21.1 (4)

Practice	Details of practice	Reason for practice	Possible adverse side effect(s)
Removal of competitors from area where crop is being grown	**Use of herbicide spray**	**To prevent weeds using essential resources such as minerals intended for the crop**	Reduction in biodiversity
Removal of insects and other pests that feed on the crop	**Use of pesticide spray**	To prevent energy present in food produced by the crop being transferred to unwanted consumers	**Reduction in biodiversity and poisoning of helpful insects**
Farm animals kept indoors	'Battery' farming of animals in confined space	**To reduce the amount of energy lost by animals and make more available for growth**	Increase in risk of disease and decrease in quality of animals' life

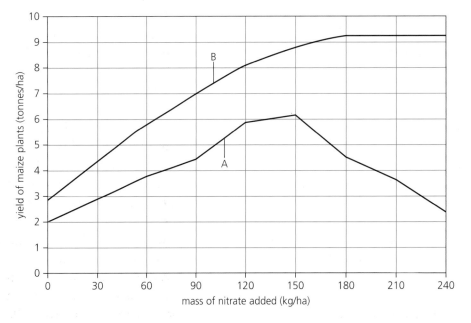

Figure An 21.1

b) 4.4 tonnes/ha (1)

c) 2.9 tonnes/ha (1)

d) i) 2.0 tonnes/ha
   ii) The maize plants obtained nitrate and potassium present naturally in the soil. (2)

e) 180 kg/ha (1)

f) Because in the absence of potassium, this is the optimum level. Higher concentrations of nitrate on its own produce lower yields of maize. (1)

3 a) i) June and September
     ii) September
     iii) June (4)

   b) i) D
     ii) During August there is no requirement for nitrate by the crop so added nitrate fertiliser would remain unused and be leached out of the soil. (2)

   c) i) A
     ii) B is wrong because the fertiliser would remain unused by the plants and be washed out by rain. C is wrong because this would be a wasteful method as the plants can only use a certain amount of fertiliser at a time. D is wrong because the fertiliser could be easily leached out of the soil before being used by the plants in spring. (4)

4 a) i) The more animal material (especially fish) that the bird eats, the higher the concentration of pesticide residue in its muscle tissues.
     ii) Animals such as fish are situated further along the food chain than plants or invertebrates and therefore eat food which has already concentrated pesticide at several links in the chain. (3)

   b) i) The sea
     ii) The sea dilutes the concentration of pesticide arriving in rivers before it enters the marine food chain. (2)

5 a) As the concentration of pesticide increases, the percentage thickness of the shell decreases. (1)

   b) i) 15%
     ii) 25
     iii) Zero success: all shells would be so thin that they would break during incubation. (4)

6 a) i) 4 km
     ii) 20 km (2)

   b) i) To increase the reliability of the results
     ii) Increases

iii) Many species of lichen are sensitive to sulphur dioxide. The further away from the power station, the lower the concentration of sulphur dioxide in the air and the higher the number of different lichen species that can survive. (4)

   c) i) 45
     ii) From NW
     iii) From SW (3)

   d) i) The increase in biodiversity of lichen species along the SE line is much greater than that along the NE line.
     ii) This is probably due to the fact that on 250 out of 365 days each year, sulphur dioxide from the power station is blown along the NE line, preventing many of the lichen species from surviving. (3)

7 a) i) Pipe W
     ii) The indicator species at site 2 is rat-tailed maggot which thrives in badly polluted water short of oxygen.
     iii) It could have been improved so that it was able to treat all the sewage before the liquid was discharged into the river. (3)

   b) i) At site 5 where the indicator species had been stonefly nymph, it now became bloodworm.
     ii) The oxygen concentration of the water at site 5 had greatly decreased. (2)

8 a) Orange tree ⟶ scale insect ⟶ ladybird (1)

   b) i) Y = predator
     ii) X = prey
     iii) Following the arrival of Y, the numbers of X dropped immediately. (3)

   c) 12 (1)

   d) 1889 (1)

   e) There were too few prey left to feed them all. (1)

   f) As the prey numbers increased, so too did the predator numbers after a short time. This caused the prey numbers to fall which in turn caused the predators to decrease in number and so on with the predators curve always tracking that of the prey. (2)

   g) i) Most of the scale insects died so the crop of fruit was very good.
     ii) A new epidemic of scale insects soon followed.
     iii) Reintroduction of ladybirds (3)

   h) The same pattern as between 1900 and 1940. (1)